APPIL

Unit
F F



# Forces and Fields

John Murray
in association with
Inner London Education Authority

**APPIL First Edition Project Team**
John Bausor (Director)
Leslie Beckett
Allan Covell
David Davies
Martin Hollins

**APPIL Revised Edition**
Co-ordinated by Martin Hollins
This unit has been revised by Johnathan Osborne and edited by Allan Covell

Design, typesetting and page layout by Tony Langham
Diagrams by Technical Art Services and John Sangwin



Published by John Murray (Publishers) Ltd,
50 Albemarle Street, London W1X 4BD

APPIL first edition 1979–80
This edition 1989
Reprinted 1991



Printed in Great Britain by St Edmundsbury Press Limited

*British Library Cataloguing in Publication Data*

Forces and fields.
Student's unit.
1. Electric fields & magnetic fields – For schools
I. ILEA Physics Project Team II. Series
531.1′41

ISBN 0-7195-4577-3

# Contents

# How to use APPIL

APPIL is a programme for independent learning. This unit is not a textbook: it is a guide to using texts, experiments and other resources to help you to learn about the topics in your A-level physics syllabus.

There are sections of text in this unit which are to be read as in any other book, but much of the guide is concerned with helping you through other activities designed to produce effective learning when you work independently. For a fuller explanation of the way APPIL is written you should read the *Student's resource book*. What follows is a brief summary.

### Objectives

What is to be learnt is stated at the beginning of each topic - a general statement of what you will be doing, and more detailed objectives to be achieved. The objectives are particularly important, because they tell you what you should be able to do when you have finished working through the topic, and so give you extra help in organising your learning. You will probably wish to refer to them when you have finished each topic.

### Experiments

> **E Experiment FM 12**
> **Barton's pendulums**
>
> The aim of this experiment is to observe what happens when a system is made to vibrate at some frequency other than its own natural frequency of vibration.

These are a very important part of the course. Each experiment is referred to at the most appropriate time in the text. You should aim to organise your work so that it can be done at that time. Most experiments should not take longer than an hour if the apparatus is available. The experiments are listed at the beginning of each topic, with an estimate of the time required and notes of any other factors to be taken account of in planning. Each experiment contains questions which require written answers. In general, there is no value in copying out the instructions given, but notes on any details which might be useful for revision should be made.

### Self-assessment questions

> **Q 5.19 Self-assessment question**
>
> A hi-fi record-player turntable has a mass of 1.28 kg and a radius of gyration about its centre of 25 cm. What is its moment of inertia? What torque will be required to accelerate it up to 33 r.p.m. in 1 second? ▪

These test your understanding of the work you have done, and will help you to check your progress. They are not intended to be difficult: you should be able to answer most of them quite easily.

The answers to self-assessment questions are given at the end of the book, but if you look at the answer before you have tried the question you will not be involved in the learning process and your learning may suffer.

▪ indicates the end of a question.

### Development questions

These are included to involve you in a proof or idea which is being developed in the text. The answer to a development question is usually in the text, but involving yourself in the development helps you to learn: just looking at the answer is not so effective. If the answer is given at the end of the book the question is marked with an asterisk*.

### Discussion questions

> **Q 5.33 Discussion question**
>
> Some small toy cars use for propulsion the energy stored in a flywheel; these toys are popularly known as 'friction-powered'. Why? What would be a better name?
>
> How does a yo-yo work?
>
> Can you find any other examples of toys using 'rotatory stores' of energy? How about designing one? ▪

These are questions where there is no 'right answer'. Often they cover technical applications, problem-solving and social issues. It is suggested that you do some background research and reading - or just thinking, to answer these.

### Study questions

For these you will need to use resources apart from this unit: for example, textbooks or experimental results. For all questions, general references are given to basic books at the start of each topic. It is not expected that you will be able to consult all the references given but you should always use more than one when possible.

This type of question usually requires longer answers than the others. These answers, in many cases, form a basis for your notes for the final examination and are therefore very important. Full answers are not usually given in this guide, though hints and partial answers are sometimes given. These questions are marked with an asterisk. Your answers to study questions should be handed in regularly for marking.

## Questions on objectives

> 1 A particle moves with a uniform angular velocity $\omega$ in a circular path of radius $r$. Show that the projection of the motion of the particle on a diameter of the circle is simple harmonic motion. *(objective 4)*

These are groups of questions which come at the end of each topic and are related to the objectives at the beginning of each topic. Answering these will help you to tell whether you have achieved the objectives. The answers are given in the *Student's resource book.*

## End-of-unit test

This is to enable your teacher to check the value of the course to you. You will be asked to do this test when you have completed the unit, and will be given details at the appropriate time.

## Examination questions

Questions in the test, and some of the other questions, are taken from past papers. Your teacher will probably set some more of these during the course, for revision purposes.

## Audio-visual aids /Computer programs

> C **Computer program**
> **Simple harmonic motion**
>
> This program enables you to explore simple harmonic motion and its relationship with uniform circular motion.

Audio-visual aids and computer programs are recommended in some topics. You should ask your teacher if these are available.

## Background reading

> **Background reading**
>
> For an exciting account of the work of Galileo and Newton, read Koestler, A. *The sleepwalkers.*

This refers to books which are useful for a more detailed study of certain topics. They are also often interesting to read in their own right, and sometimes put the physics of the syllabus in its historical, social or technological context.

# References

References are made where appropriate to textbooks. A list of the abbreviations used can be found at the end of this unit.

## Comprehension

> C **Comprehension FM 2**
> **Building the wind-up tube train**
>
> This is an article on the use of flywheels in the New York underground system.

These extracts from recent scientific and technological articles have a similar aim. They are followed by questions designed to check your understanding. They will be found in Part 4 of the *Student's resource book.*

## Extension

> **EXTENSION**
>
> Q **2.30 Study question**
>
> (a) Why is a violin box shaped the way it is? (b) How can a violinist change the quality of note produced by this instrument? (c) In what other shapes are sound boxes made? Suggest how they affect the sound produced.n

These boxed sections are included to provide:

(a) additional material of general interest and importance, or
(b) more detailed treatment of topics for more able students.

Sections marked ● are not required for some syllabuses and may provide additional useful extension material. Your teacher will advise you of the particular requirements of your syllabus.

## Organising your time

In this programme of work there is a variety of activities. Some of them, like experiments, need a laboratory, and you may also need to use the library. You must, therefore, organise your time so that you can make the best use of the resources available.

When you start a topic, look through it and see what activities are included, then allocate each activity a time on your work schedule. Make sure, for example, that you do the experiments when you are timetabled in a laboratory. Follow the sequence in this guide if you can, but this may not always be possible.

The recommended time for completion of the work in each topic is given at the beginning of each topic. This assumes that you spend 8-10 hours each week on physics, divided between class time, private study and home study. It is important to try to complete the unit in a set time. You should ask your teacher for a **progress monitor**. It will help you to plan your time.

APPIL FF

# Introduction to the unit

In this unit you will be studying an important concept in physics - a field of force - by considering gravitational, electric and magnetic fields. In most cases, the topics are covered to A-level standard: others will be developed in a later unit, and there is an indication at appropriate points in the text.

**Topic 1** opens with a consideration of gravitational forces and how they account for the observed motion of planets and satellites. Information from the Apollo 11 space flight is used to verify Newton's prediction of an inverse square law of gravitational force.

Results from experiments with charged spheres lead to a similar relationship for electric forces.

**Topic 2** considers the way in which a scientist analyses the forces between masses and between like and unlike charges by introducing the idea of a field. The way in which a field can be described and its strength measured is studied in detail. The topic concludes with a discussion of the electric field around different shaped conductors, considers some important industrial applications in electrostatics and the Van de Graaff generator.

**Topic 3** is concerned with magnetic fields. How such a field can be represented and its strength measured, and how it interacts with moving charges are studied in detail. The magnetic field due to different shapes of conductors is investigated using a Hall probe and equations which relate the magnetic field to the current and geometry of the conductors are derived. The topic concludes with a development of the principles of an important measuring instrument, the moving-coil galvanometer.

**Topic 4** is about motion in gravitational, electric and magnetic fields. In section 4.1 the motion of satellites in a circular orbit and the conditions under which a body can escape from a gravitational field are studied. In section 4.2 the forces acting on charged particles in electric and magnetic fields are studied. The ratio of charge to mass for electrons is measured. These ideas are then used to explain how electron beams are accelerated, focused and deflected in an oscilloscope and how high energy charged particles are produced in a cyclotron.

**Recommended study times**

You should spend about 6 weeks on this unit as follows:

| | |
|---|---|
| Topic 1 | 1.0 week |
| Topic 2 | 2.0 weeks |
| Topic 3 | 1.5 weeks |
| Topic 4 | 1.0 week |

# TOPIC 1

# Gravitational and electric forces

**Study time: 1 week**

## Summary

The satellite falls and orbits the earth, responding to its gravitational field. Electric fields are less obvious (though they hold molecules together and drive currents). The platypus provides an example of a biological system that detects the electric field of its underground prey. We shall study simpler systems to make clear how fields act in space.

## Objectives

When you have completed the work in this topic you should be able to:

1 Use correctly the following scientific terms:
conductor, insulator, unlike and like charges, coulomb, permittivity, electrostatic induction.

2 Define and derive the dimensions for the universal gravitational constant, *G*.

3 Recall the standard symbols and units for:
charge, gravitational constant, permittivity constant $\varepsilon_0$.

4 State how the force of gravitational attraction between two particles depends upon the mass of the particles and their distance apart.

5 State how the force of electric attraction or repulsion between two charged particles in a vacuum depends upon the magnitude of the charges and their distance apart.

6 Solve problems involving gravitational and electric forces.

7 EXTENSION
Describe a laboratory experiment to determine the gravitational constant.

8 EXTENSION
State and explain Kepler's laws of planetary motion.

## Experiments

FF1 The force between two charged spheres (1 hour)

## References

| | |
|---|---|
| Akrill | Chapters 12, 14.3, 15.1 |
| Bolton | Chapters 12 and 13 |
| Duncan | Chapters 8 and 13 |
| Muncaster | Chapters 8 and 39 |
| Nelkon | Chapters 2 and 6 |
| Wenham | Chapters 28, 29, 30 |
| Whelan | Chapters 20 and 42 |

## 1.1 Gravitational forces

Gravitational forces are, perhaps, more familiar to us than electrical ones. The opening photograph shows a satellite orbiting the earth held at a particular radius by gravity.

### Q 1.1 Discussion question*

**Figure 1.1**

Investigators have shown that a duck-billed platypus (figure 1.1) is able to detect electrically a buried 1.5 V cell.

What purpose do you think this serves? ■

### Q 1.2 Self-assessment question

(a) At the earth's surface freely falling bodies have an acceleration of 10 m s$^{-2}$. What is the gravitational force per kilogram?
(b) Calculate the gravitational force which is exerted on a ball of mass 5 x 10$^{-2}$ kg.
(c) What is the force which the ball exerts on the earth?
(d) Why doesn't the earth appear to accelerate towards the ball? Explain your answer. ■

This attractive force is called a **gravitational** force or the force of gravity. It also accounts for the observed motion of the planets and the behaviour of satellites. The force of attraction exerted by the earth on the satellite provides the necessary centripetal force to keep it in its orbit.

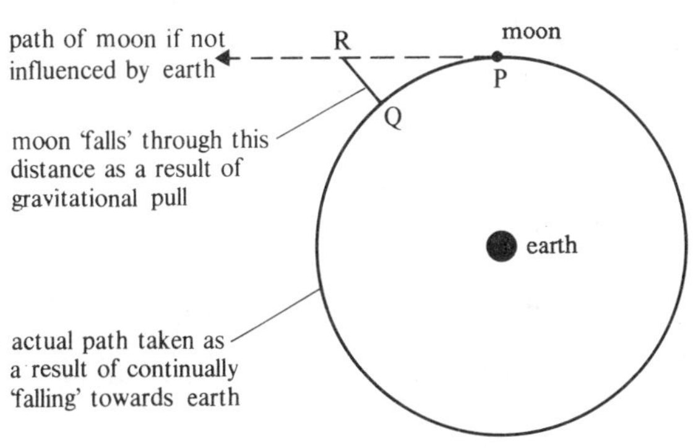


**Figure 1.2**

This explanation was first put forward by Sir Isaac Newton in a book called *Principia Mathematica* - the mathematical principles of natural philosophy. In it he proposed a precise law of attraction to explain the motion of the moon around the earth and the planets around the sun (figure 1.2 ), and suggested that the same law applies also to objects near the earth. At the time, this was a strange idea: people thought 'earthly' and 'heavenly' objects were quite different.

Newton stated that all particles of matter attract each other, and that the force between them depends on their masses and their distance apart. Consider two particles A and B, masses $m_1$ and $m_2$ respectively, separated by a distance $r$ (figure 1.3). A particle is a convenient imaginary object, which has mass but no size. Matter can be thought of as made of particles at suitable distances apart.

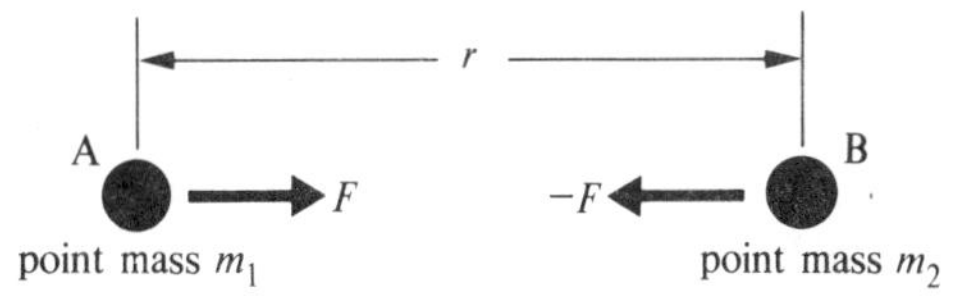


**Figure 1.3**

Particle A exerts a force $F$ on particle B, and B exerts equal and opposite force on A. According to Newton the magnitude of this force is
(i) directly proportional to the product of the masses of the two particles

$$F \propto m_1 m_2$$

(ii) inversely proportional to the square of their distance apart.

$$F \propto \frac{1}{r^2} \ (m_1 \text{ and } m_2 \text{ constant})$$

Combining these statements, the gravitational attraction $F$ between two particles of masses $m_1$ and $m_2$ separated by a distance $r$ is given by

$$F \propto \frac{m_1 m_2}{r^2}$$

or

$$F = \frac{G m_1 m_2}{r^2}$$

where $G$ is a universal constant whose value must be found experimentally.

*Note*: Newton showed that from a point outside a sphere the sphere appears to behave as if all its mass is concentrated at its centre. He delayed the publication of his work on gravitation for over twenty years until he had proved this. Thus, we can apply his law of gravitation to bodies such as the earth and moon, and $r$ will be the distance between the centres of the masses.

### Q 1.3 Self-assessment question

Suggest why Newton assumed that the gravitational force

was

(a) proportional to the mass of the attracted body,
(b) proportional to the mass of the attracting body. ■

## An inverse square law

About a century before Newton, over a period of twenty years, the Danish astronomer Tycho Brahe made accurate observations on the motion of the planets. This data was studied by Kepler, who put forward a description of planetary motion which is summarised in three laws. These laws describe how the planets behave in the solar system. They do not indicate why a planet moves in an elliptical orbit, or what provides the central force, that is, the force towards a focus of the ellipse.

Kepler's laws provided Newton with a summary of the observed motion of the planets. From these he was able to show that the force on the planet must be directed towards the sun and that it must be inversely proportional to the square of the distance between the sun and the planet (assuming a circular orbit).

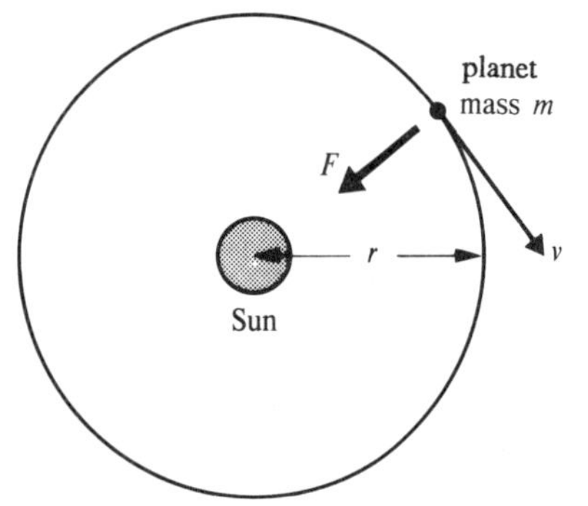


**Figure 1.4**

### Q 1.4 Development question

For a circular orbit we can express Kepler's third law as

$$T^2 \propto r^3$$

where $T$ is the period of revolution and $r$ is the radius of the circle. We will now show how this leads to an inverse square law of attraction.

Suppose that a planet of mass $m$ moves with a speed $v$ in a circular orbit of radius $r$ (figure 1.4).

(a) The acceleration of the planet towards the centre of the orbit is $v^2/r$. Show that the force which is necessary to cause the planet to move in this orbit is given by

$$F = \frac{mv^2}{r}$$

(b) If $T$ is the time taken for the planet to make one complete revolution show that

$$v = \frac{2\pi r}{T}$$

(c) Hence show that

$$F \propto \frac{1}{r^2}$$

The law of gravitation which Newton developed to explain the motion of the planets has in recent times been used to calculate and predict the paths of satellites and lunar probes.

---

### C Comprehension FF1
**Data analysis exercises**

Part A presents data obtained by Kepler and asks you to check the relationship between radius and period of planetary orbits.
Part B presents data from the Apollo 11 space flight, which covers a period when the spacecraft was travelling more or less directly away from the centre of the earth, with the motors shut down. This data can be used to support the inverse square law of attraction.

---

**EXTENSION**

### Q 1.5 Study question

Explain with the aid of figure 1.2 how Newton tested the universality of the law of gravitation by comparing the gravitational effect of the earth on the moon with the effect of a body falling near the earth's surface. ■

### Kepler's laws

### Q 1.6 Study question

State and explain Kepler's laws of planetary motion. ■

## 1.2 Universal gravitational constant

The constant $G$ is called the universal constant of gravitation; it has a magnitude of 6.7 x $10^{-11}$ N m$^2$ kg$^{-2}$.

Rearranging the equation $F = G\, m_1 m_2 / r^2$

we have

$$G = \frac{Fr^2}{m_1 m_2}$$

From this relationship we can see that $G$ is numerically equal to $F$ when $r$, $m_1$ and $m_2$ are all equal to unity. $G$ is numerically equal to the gravitational attraction between two point or spherical 1 kg masses when they are 1 m apart. This is a very small force indeed!

### Q 1.7 Self-assessment question

(a) How does this force compare with the earth's gravitational pull on a mass of 1 kg?
(b) Show that the dimensions of $G$ are $M^{-1} L^3 T^{-2}$. ■

### Q 1.8 Discussion question*

(a) What do you think is meant by a universal constant?
(b) Can you think of any situation or experimental result that would lead us to doubt the constancy of $G$? ■

## Q 1.9 Self-assessment question

(a) A communications satellite has a weight $W$ when on the surface of the earth. What will be the earth's gravitational pull on the satellite when it is in a circular orbit of radius twice the radius of the earth?
(b) It is more realistic to consider the satellite in an orbit of radius 1.1 times the earth's radius (i.e. about 640 km above the earth). What is the gravitational pull in this case? ■

## Q 1.10 Self-assessment question

Two spherical bodies A and B of the same uniform density, situated in outer space, attract each other with force $F$. What is the attraction between two similar bodies of the same density but each of twice the diameter and situated three times as far apart as A and B? ■

## Q 1.11 Self-assessment question

Estimate the force of gravitational attraction between two adult human beings when standing side by side. ■

### EXTENSION

### Determination of *G*

There have been many experiments to determine $G$. Early methods were concerned with measuring the deviation of a pendulum bob line from the vertical due to the presence of a large mountain. The first laboratory method was carried out by Cavendish in 1798 who used a torsion balance to measure the force of attraction between two small masses. His apparatus was modified by Boys in 1895 who used a quartz fibre suspension system. The quartz fibre was capable of supporting a considrable mass and also offered less resistance to twist than a wire of the same strength. This enabled him to reduce the scale of the apparatus and consequently he avoided trouble due to draughts.

## Q 1.12 Development question*

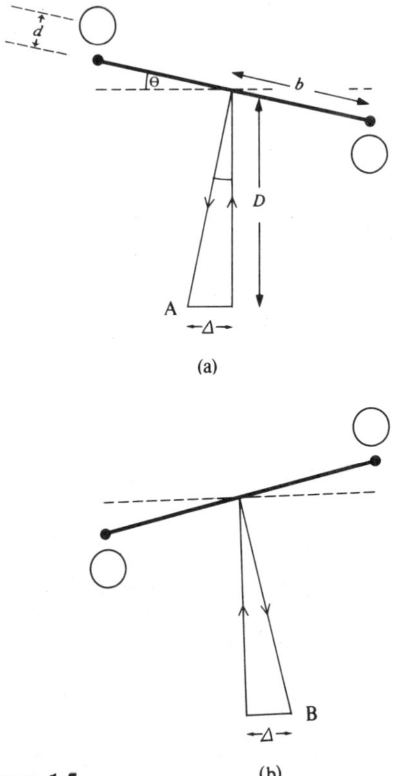


**Figure 1.5**

This question is based on a video which shows the principle of the torsion balance method for determining $G$, using film of an experiment in progress. Ask your teacher whether a copy of the video is available for you to watch. If it is not available, then obtain the relevant experimental data.

The large masses are first placed in position, as shown in figure 1.5a.

After several hours the torsion bar comes to rest and the position A of the light beam is noted.

The masses are then moved to the position shown in figure 1.5b. When the torsion bar again comes to rest, the position B of the light beam is noted.

(a) What is the difference between the two scale readings?
(b) Show that the deflection $\theta$ is given by

$$\theta = \frac{\Delta}{2D}$$

(c) What is gravitational force between the large mass $M$ and small mass $m$?
(d) What is the gravitational couple on the beam?

The gravitational couple is equal to the torsional couple $c\theta$ where $c$ is the torsional constant of the wire.

The torsion constant can be determined by finding the period of the torsion balance, since

$$T = 2\pi\sqrt{\frac{I}{c}}$$

where $I$ is the moment of inertia of the balance, equal to $2mb^2$.

The video asks you to record the data and plot the displacement time curve to obtain $T$.

(e) Use this information to calculate a value for $G$.

The constants for the apparatus are:

(i) M = 1.01 kg; mercury; outside diameter 53.0 x $10^{-3}$ m

continued

(ii) $M$ = 1.500 kg; lead; outside diameter 62.5 x $10^{-3}$ m
(iii) $M$ = 2.96 kg; lead; outside diameter 78.0 x $10^{-3}$

Small spheres $m$ = 15.0 x $10^{-3}$ kg

Distance $d$ between small and big sphere equals radius of big sphere plus 15.5 x $10^{-3}$ m

Distance $2b$ = 100 x $10^{-3}$ m $D$ = 5 m ■

The history of the development of the theory of planetary motion provides an excellent example of the growth and use of theory in science.

**Background reading**

You should consult one or more of the following references:


Feynman, R. *The character of physical law.* Chapter 1 The law of gravitation, an example of a physical law. BBC, 1965.
Gamow, G. *Gravity*, Heinemann Science Study Series no. 17, 1961.
Rogers, E.M. *Physics for the inquiring mind.* Part 2 Astronomy: a history of theory. OUP, 1960.


**Q 1.13 Study question**

Discuss the contributions of Copernicus, Kepler and Newton to the study of gravitation, and explain how the theory of universal gravitation was established. ■

### Is gravity getting weaker?

Einstein's general theory of relativity, and theories which predict that the force of gravity diminishes as the universe expands, are discussed in an article in Scientific American.

**Background reading**


Van Flandern, T.C. 'Is gravity getting weaker?' *Scientific American*, February 1976.


**C Comprehension FF2**
**Einstein's explanation for gravity**

A student, observing ants running over the surface of an apple, is led to an understanding of Einstein's modification of Newton's theory, involving the interaction of matter and spacetime.

## 1.3 Electric forces

In the previous section you learned that there is a force of gravitational attraction between all bodies. This has practical consequences only if at least one of the attracting bodies has enormous mass, as has the earth. Gravitational forces are not the only forces acting at a distance between material bodies. You have no doubt discovered that when two strips of polythene are rubbed with a cloth they exert forces on each other. These forces are called **electrostatic** or **electric** forces. To explain this we say that the materials have acquired a charge. Experiments show that there are two types of charge, one positive and the other negative, and that like charges repel and unlike charges attract.

**Q 1.14 Study question**

What is the evidence that there are *only two* types of charge?

### The electron model

All materials are atomic in structure. Each atom can be considered as a neutral, uncharged body, but it consists of a positively charged nucleus surrounded by negatively charged electrons. The electrons are held near the nucleus by electrostatic attraction. In certain substances it is possible to remove some of the surface electrons, leaving a positively charged surface on the material and taking away an equal negative charge.

**Q 1.15 Study question**

How does the electron model explain

(a) the fact that when a polythene rod is rubbed with a cloth it acquires a negative charge,
(b) the fact that when cellulose acetate is rubbed with a cloth it acquires a positive charge? ■

### Conductors and insulators

Experiments show that we can distinguish between two types of materials
(i) those that retain an electric charge on the surface - called insulators or non-conductors (such as polythene, glass, etc.),
(ii) those that allow charges to be placed on the surface which, when isolated, repel each other to redistribute themselves over the surface - called conductors (such as metals or carbon).

We can also distinguish between conductors and insulators by their ability to conduct an electric current.

**Q 1.16 Study question**

Explain how the electron model can account for the difference between conductors and insulators. ■

### Charging by induction

Figure 1.6 illustrates how a metal sphere can be charged by a process known as induction. The sequence of events is as follows:

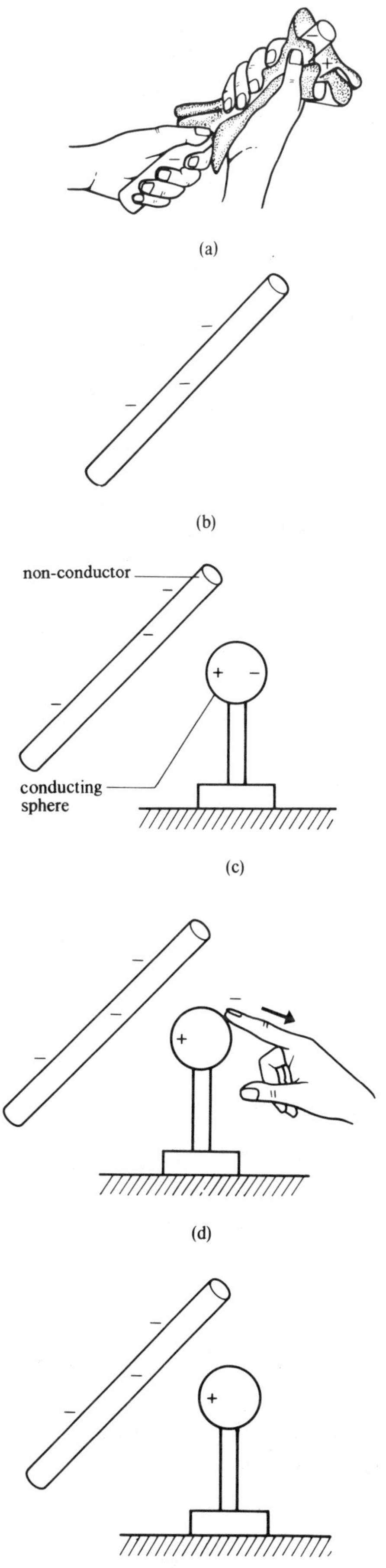


**Figure 1.6**

(a) When the rod of polythene is rubbed with a cloth, the cloth loses electrons to the polythene; the cloth is charged positively, the rod negatively.
(b) When the cloth is removed the rod remains charged.
(c) When a conducting metal sphere on an insulating stand is brought near the rod, some of the conduction electrons are repelled to one side leaving the sphere with different charges on the two sides. This separation of charges by an electric force is known as **induction**.
(d) The sphere is now connected to earth by touching it.
(e) Finally, the polythene rod is removed and the metal sphere has a positive charge.

## Q 1.17 Study question

(a) How does the electron model explain the fact that the metal sphere has a positive charge?
(b) Draw a series of diagrams to illustrate how, starting with a charged acetate rod, the sphere is left with a negative charge. ■

The charge that is left is called the induced charge and it has the *opposite* charge to the inducing charge.

## Q 1.18 Discussion question

How do you account for the following observations?

(a) When somebody walks across a carpet and then touches a radiator, a spark may be produced.
(b) When a plastic comb or ruler is rubbed, it can attract small pieces of paper. ■

# The coulombmeter

How can we tell whether a body has a positive or negative charge? We certainly can't decide by just looking at it! (Or by weighing - the mass of the displaced electrons is infinitesimally small compared with that of the body itself.) One possible way is to explore the region around a charged body with a small charged sphere, e.g. a pithball or polystyrene sphere suspended by nylon or silk thread.

## Q 1.19 Self-assessment question

What observations would indicate the presence of

(a) a positively charged body, and
(b) a negatively charged body? ■

This, however, is not a very convenient method; we need a better instrument - the coulombmeter is one.

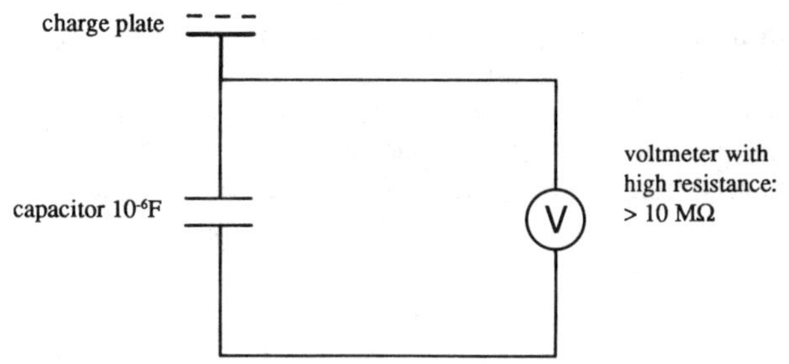


**Figure 1.7**

## Q 1.20 Study question

Figure 1.7 shows the basic design of a coulombmeter.

(a) What is the time constant for this circuit?
(b) Explain how this device measures charge.
(c) How would it show the presence of positive or negative charge?
(d) The gold leaf electroscope can be adapted to measure charge. How it it similar to, and different from, the coulombmeter? ■

## 1.4 Coulomb's law

The magnitude of the force between two 'point' charges was investigated in 1785 by the French physicist Coulomb. He used a torsion balance that he had invented (similar to the one shown in figure 1.8).

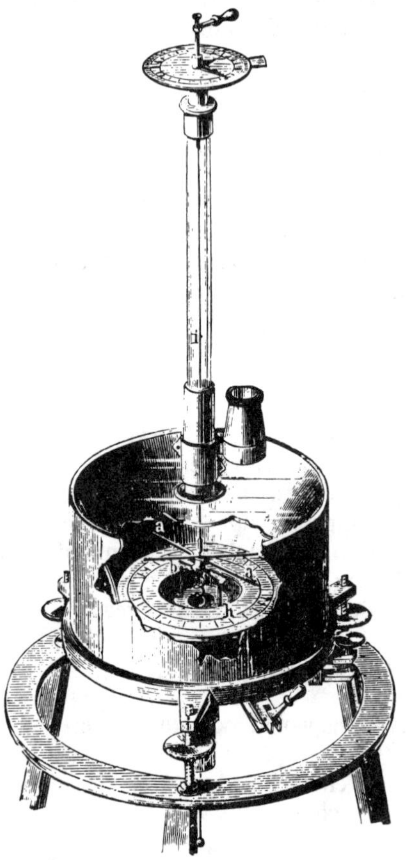

Figure 1.8

An insulated rod, which was suspended by means of silver wire, twisted when a force was exerted on one end. This twisting effect was used to measure the force between a charged body attached to one end of the rod and another charged body placed near it. (Compare this apparatus with that used by Cavendish for the determination of *G*.)

In the following experiment you will set up a simplified version of Coulomb's apparatus.

## E Experiment FF1
**The force between two charged spheres**

The aim of this experiment is to investigate how the force between two charged spheres depends upon the magnitude of the charges and their distance apart.

### An inverse square law

Suppose there are two point charges A and B, charges $Q_1$ and $Q_2$ respectively, separated by a distance *r*. (The letter *Q* is used to symbolise the quantity of electricity, i.e. the charge.)

Consider the force *F* on B (figure 1.9a). If both charges are positive (or both negative) there will be a repulsive force on B. (B also exerts an equal and opposite force on A.)

If one of the positive charges is replaced by a negative charge, there is an attractive force between the two charges (figure 1.9b).

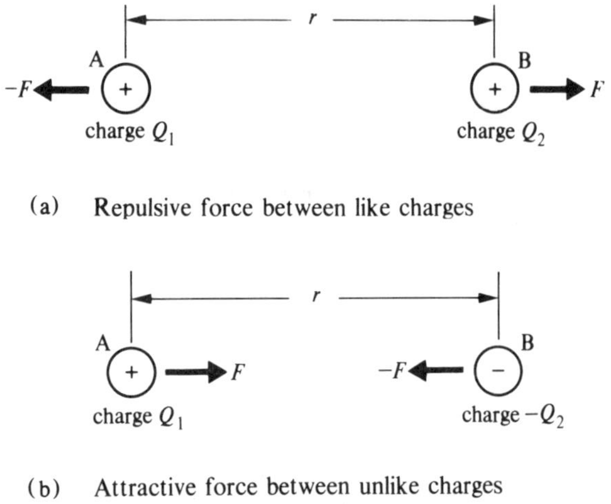


Figure 1.9

Coulomb's results showed that the magnitude of this force is (i) directly proportional to the product of the point charges

$$F \propto Q_1Q_2 \ (r \text{ constant})$$

(ii) inversely proportional to the square of their distance apart

$$F \propto \frac{1}{r^2} \ (Q_1 \text{ and } Q_2 \text{ constant})$$

(iii) dependent upon the insulating medium which separates them.

The electric force of repulsion between two like charges, $Q_1$ and $Q_2$, separated by a distance *r* in a vacuum is given by

$$F \propto \frac{Q_1Q_2}{r^2}$$

or

$$F = \frac{kQ_1Q_2}{r^2}$$

where *k* is a constant whose value is determined by experiment.

## Q 1.21 Self-assessment question

The information in table 1.1 was obtained from an experiment with charged spheres. The deflection of the suspended sphere from its mean position was $x$, and the separation of the spheres was $d$. The suspended sphere hung 0.5 m below its support, as shown in figure 1.10a.

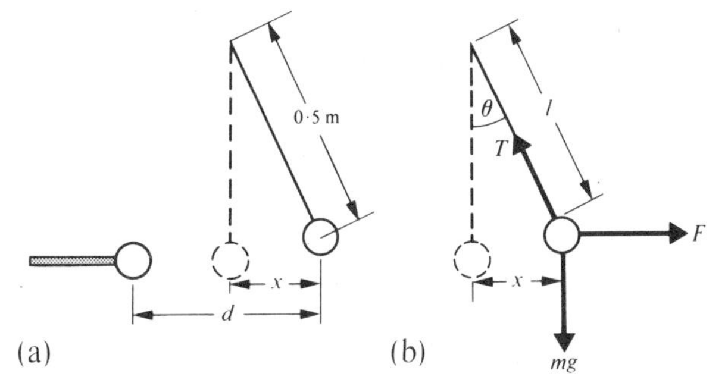


**Figure 1.10**

**Table 1.1**

| Deflection $x/10^{-2}$ m | Separation $d/10^{-2}$ m | (Separation)$^2$ $d^2/10^{-4}$ m$^2$ | 1/(Separation)$^2$ $d^{-2}/10^4$ m$^{-2}$ |
|---|---|---|---|
| 1.0 | 11.0 | 121.00 | 0.008 |
| 3.5 | 6.5 | 42.25 | 0.024 |
| 5.5 | 5.5 | 30.25 | 0.033 |
| 7.5 | 4.5 | 20.25 | 0.049 |
| 9.0 | 4.0 | 16.00 | 0.062 |
| 11.0 | 3.5 | 12.25 | 0.082 |

From figure 1.10b we have

$$F = T \sin \theta \text{ (resolving vertically)}$$

$$mg = T \cos \theta \text{ (resolving horizontally)}$$

Thus

$$F = mg \tan \theta$$
$$= mgx$$

(a) Plot a graph to test whether the repulsive force $F$ on the suspended sphere is proportional to $1/d^2$.
(b) Write down an expression for the slope of the graph in terms of $k$, $Q_1$, $Q_2$, $l$, $m$ and $g$.
The charge on each sphere was 5 x $10^{-9}$ C and the mass of the suspended sphere was 1.1 x $10^{-4}$ kg.
(c) Estimate a value for the force constant $k$. ■

## The unit of charge

Coulomb's law can be used to define a unit of charge. We could decide to make the magnitude of $k$ exactly 1, and define a unit of charge so that two unit charges separated by a unit distance *in vacuo* exert unit force on each other. There exists a unit of charge based on this choice; it is called the electrostatic unit of charge or statcoulomb. However, in the international system of units (SI) that we use, the unit of charge is derived from the unit of current, the ampere. The unit of charge is the coulomb and is defined as the charge that flows past a given cross-section in a conductor in one second when the current is equal to one ampere. If the coulomb is adopted as the unit of charge, then the constant $k$ in Coulomb's law must be found by experiment, e.g. by measuring the force between known point (or spherical) charges separated by a known distance. (Compare this with the determination of $G$.) This is not the method that is adopted in practice (see unit *Electronic systems*).

## The permittivity constant

The value of the force constant $k$ for a vacuum is approximately 9.0 x $10^9$ N m$^2$ C$^{-2}$.

It is, however, more usual to write the constant in the form

$$k = \frac{1}{4\pi\varepsilon_0}$$

where $\varepsilon_0$ (pronounced epsilon nought) is called the **permittivity constant** or the **permittivity of free space.**

Thus, for a vacuum, we can write Coulomb's law as

$$F = \frac{1}{4\pi\varepsilon_0} \cdot \frac{Q_1 Q_2}{r^2}$$

*Notes*: (i) The reason for introducing the constant $1/4\pi$ is not immediately obvious but, by doing this, in equations which refer to situations where there is no spherical symmetry (for example, a uniform field) does not appear. Equations which are written in this way are said to be rationalised.
(ii) It is usual to assume that the permittivity of air is the same as that of a vacuum.
(iii) The usual unit for permittivity is the farad per metre, F m$^{-1}$. This will be introduced in the unit *Electronic systems.*

## Q 1.22 Self-assessment question

(a) Calculate the force of attraction between an electron and a proton which are $10^{-8}$ m apart.
(b) If the electron and proton were free to move towards each other, what would be the acceleration of each?
(charge on an electron = -1.6 x $10^{-19}$ C
mass of electron = 9.1 x $10^{-31}$ kg
charge on a proton = 1.6 x $10^{-19}$ C
mass of proton = 1.7 x $10^{-27}$ kg

$$\frac{1}{4\pi\varepsilon_0} = 9.0 \times 10^9 \text{ N m}^2 \text{ C}^{-2}$$

## Forces in the nucleus

In the unit *Electrons and the nucleus* evidence is given for a model of the atom which consists of a central nucleus, of diameter about $10^{-15}$ m, surrounded by electrons. The nucleus consists of roughly equal numbers of positively charged particles, protons, and particles with no charge, neutrons.

If like charges repel, why don't the protons fly apart? Is it because of gravity?

**Q 1.23 Self-assessment question**

Find the ratio of the electric and gravitational forces between two protons which are placed a distance *d* apart. ■

Protons are not held together by gravitational attraction, because this force is approximately $10^{36}$ times smaller than the repulsive force between the protons.

Clearly, there must be some other force which holds the nucleus together. This force, called the nuclear force, has a very short range. Nuclear forces do not extend outside the nucleus, and they do not follow an inverse square law. They will be considered in more detail in the unit *Electrons and the nucleus.*

EXPERIMENT FF1

# The force between two charged spheres

## Apparatus

- ❍ 2 metallised polystyrene spheres
- ❍ 2 insulating rods
- ❍ metre rule
- ❍ slotted base
- ❍ stand, clamp and boss (for supporting screen and insulating rod)
- ❍ means of charging spheres: van de Graaff generator or electrophorus plate
- ❍ lamp, holder and stand
- ❍ transformer
- ❍ screen
- ❍ digital balance accurate to 0.01 g
- ❍ hair dryer

1 Set up the apparatus as shown in figure 1.11. The spheres should be positioned in front of and at the same height as the lamp so that their shadows are cast onto a sheet of graph paper which is fastened to a screen.

2 Use the tare to zero the balance.

3 Charge the spheres by contact using the plate of an electrophorus or the dome of a van de Graaff generator.

4 Bring the mounted sphere P towards sphere Q so that they are about 7 cm apart.

5 Measure and record the distance $d$ between the centres of the shadows of the two spheres. Record the reading on the balance and hence calculate the force $F$.

6 Move the sphere P, a few centimetres at a time, towards Q. Obtain a series of values for $F$ and $d$. Take the readings in quick succession. Why?

*Note.* It is a good idea to use a hair dryer to warm and dry all apparatus. Why?

7 Plot a graph that will enable you to deduce a relationship between the force and the distance between the spheres.

8 What are the main sources of error in this experiment?

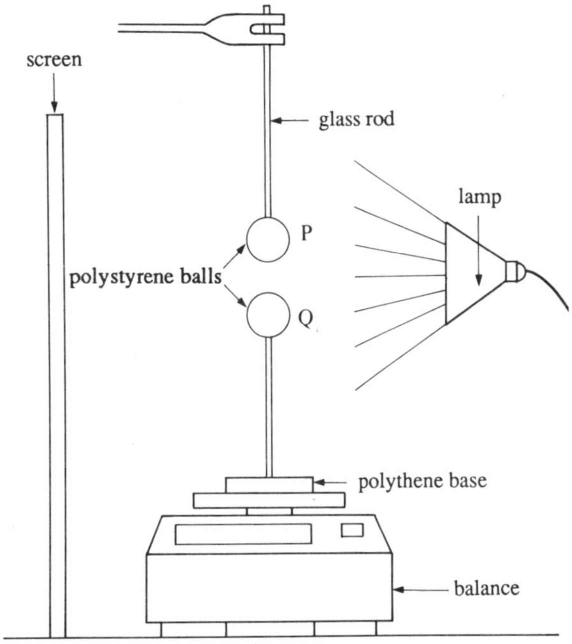


**Figure 1.11**

# Questions on objectives

Before you attempt these questions, check through the list of objectives for this topic and make sure your notes on the chapter have not left out anything important.

1 What are the dimensions of the gravitational constant $G$?

**A** $[M^{-1} L T^{-2}]$
**B** $[M^{-1} L^2 T^{-2}]$
**C** $[M^{-1} L^3 T^{-2}]$
**D** $[M^{-2} L^2 T^{-2}]$
**E** $[M^{-2} L^3 T^{-2}]$ *(objectives 1 and 2)*

2 Show that the maximum gravitational force between two identical spheres is proportional to the fourth power of their radius. State the assumptions you make in order to perform the calculation. *(objectives 4 and 7)*

3 Make an order-of-magnitude calculation of the force of electrostatic attraction between two adults standing side by side if $2.0 \times 10^{-6}$ C of charge has been transferred from one person to the other.

$$\frac{1}{4\pi\varepsilon_0} = 9.0 \times 10^9 \text{ N m}^2 \text{ C}^{-2}$$

*(objectives 5 and 7)*

4 When two small charged objects are placed a certain distance apart the electrostatic force between them is found to be 3.0 N.

(a) What would the force become if the separation were halved?
(b) What would the force be if both charges were doubled, the separation remaining as it was originally?
*(objective 5)*

5 Figure 1.12 shows a plan view of the apparatus for a laboratory determination of $G$. A and B are two small spheres, each of mass $m$, attached to a light rod of length $2a$. The rod is suspended from a torsion fibre, which is such that a couple $c$ applied to the rod produces an angular deflection of one radian. The larger spheres D and E, each of mass $M$, are placed close to A and B. When the rod settles it is deflected through an angle $\theta$, so that the distance between the centres of A and D, and B and E, is $d$.
(a) What is the torque exerted by the torsion fibre when the rod is in its deflected position?
(b) Find an expression, in terms of $M$, $m$, $a$, $d$, and $G$, for the couple due to the gravitational attractions between A and D, and between B and E.
(c) Derive an expression for $G$, assuming that the forces between A and E, and between B and D, can be neglected.
(d) Will the assumptions made in part (c) give a value of $G$ which is too large, or too small? Explain your answer.
*(objective 6)*

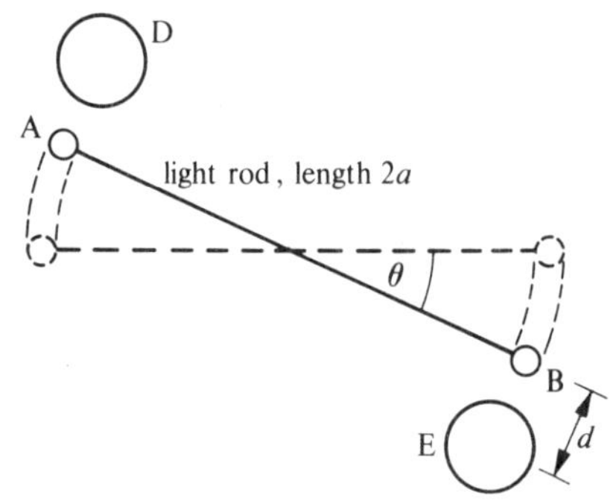


**Figure 1.12**

TOPIC 2

# Field and potential

**Study time: 2 weeks**

## Summary

This topic develops the concepts of a field of force, and of the potential at a point in such a field. You will be studying both gravitational and electric fields, their similarities and differences, and the ways in which we can represent them, mathematically and diagrammatically.

The photograph shows a phenomenon which results from electric fields in the atmosphere. The earth has a negative charge - which way does lightning travel?

## Objectives

When you have completed the work in this chapter you should be able to:

1 Use the following scientific terms correctly:
electric field, gravitational field, electric field line, gravitational field line, equipotential surface.

2 Define and use the following scientific terms, and recall their standard symbols and units:
electric field strength, gravitational field strength, electric potential, gravitational potential, potential difference, potential gradient, electronvolt.

3 Describe, with the aid of sketches, the general shapes and relative intensities of the electric and gravitational fields produced by various arrangements of charges or masses.

4 Derive a relationship between gravitational field strength and gravitational constant.

5 Derive relationships between electric field strength and potential, and between gravitational field strength and potential.

6 Recall and derive the formulae for the gravitational potential and electric potential at a point due to an isolated point mass and charge respectively.

7 Explain, with the aid of suitable graphs, how gravitational field strength and potential, and electric field strength and potential, vary with distance from a point mass or charge respectively.

8 Describe the similarities and differences between electric and gravitational forces, fields and potentials.

9 Describe how charge distributes itself over the surface of a conductor.

10 Explain the discharging action of sharp points and describe two important applications of the effect.

11 Describe the construction and explain the action of a Van de Graaff generator.

12 Solve problems involving gravitational and electric field strength, and gravitational and electric potential.

## Experiments

| | | |
|---|---|---|
| FF2 | Electric fields | (1 hour ) |
| FF3 | The potential around a sphere | (0.75 hour) |
| FF4 | Relationship between charge and potential | (0.5 hour ) |

## References

| | |
|---|---|
| Akrill | Chapters 12 and 15 |
| Bolton | Chapters 12 and 13 |
| Duncan | Chapters 4, 8 and 13 |
| Muncaster | Chapters 8 and 39 |
| Nelkon | Chapter 6 |
| Wenham | Chapters 28, 29 and 30 |
| Whelan | Chapters 20, 43, 44 and 45 |

## 2.1 Fields of force

### Electric field

An electric field is the name that is given to a region in space that surrounds an electric charge, where its influence can be observed. In this region a force will be exerted on a charge which is placed in the field. This concept of a field of force, or simply a field, is a very useful model which is used to explain forces which act at a distance, that is, without any material contact.

**Q 2.1 Discussion question***

(a) What other types of force have fields?
(b) Is there any limit to the extent of a field?
(c) Is it possible to make a boundary of these fields?■

A charged body, for example a positively charged sphere A (figure 2.1), produces an electric field. If a point positive charge B is placed near to the positively charged sphere it will experience a force. We say that the electric field that is set up by the charged sphere A enables A to exert a force on the point charge B.

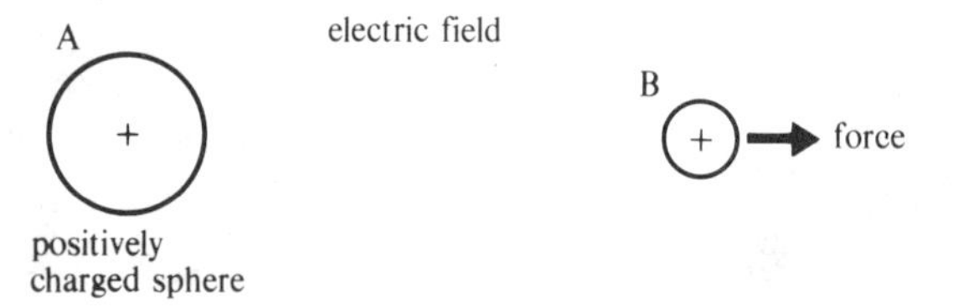


**Figure 2.1**

**Q 2.2 Self-assessment question**

(a) If B was replaced by a point negative charge C, in which direction would the force act?
(b) If the charge on sphere A was negative, what effect would this have on the positive charge B? ■

### Electric field strength

This idea of a field is extremely useful to the physicist, and to make it more meaningful it is important to have an understanding of how the magnitude of a field can be calculated. This is done by considering the effect that the field has on a very small positive charge of very small dimensions which is placed in the field.

**Q 2.3 Self-assessment question**

(a) Why must we consider a charge which has small dimensions? Why must it be a small magnitude of charge?
(b) What factors determine the magnitude of the force exerted on a small positive charge which is placed in an electric field produced by a positively charged sphere? ■

**Q 2.4 Study question**

Write a definition of the electric field strength $E$. Give the equation relating $E$ to the force exerted on a small charge $Q$ and hence a unit in which $E$ can be measured. Is $E$ a vector or a scalar quantity? ■

We will show later in this topic that an equivalent unit to the newton per coulomb, $N\ C^{-1}$, used more frequently, is the volt per metre, $V\ m^{-1}$.

**Q 2.5 Self-assessment question**

(a) If a charge of $5.0 \times 10^{-5}$ C experiences a force of 20 N, what is the electric field strength?
(b) What is the force exerted on an electron by an electric field of strength $5.0 \times 10^{4}\ N\ C^{-1}$ (magnitude of charge on electron $= 1.6 \times 10^{-19}$ C). ■

### Electric field lines

An electric field can be represented by electric field lines, sometimes referred to as lines of force. This idea helps a physicist or electrical engineer to visualise the shape of an electric field. It is possible to make electric field patterns visible in a similar way to that in which iron filings make a magnetic field pattern visible. This is achieved by using tiny needles of insulating material, such as semolina powder or grass seed, which align themselves in the direction of the electric field (figure 2.2).

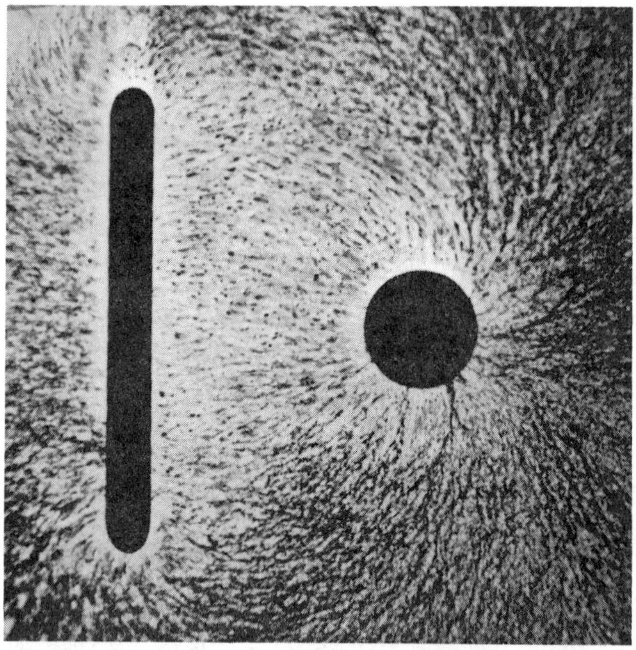

**Figure 2.2**

Electric field lines can be drawn to represent the electric field so that

(i) the tangent to the field line at a point represents the direction of the electric field at that point,
(ii) the number of field lines per unit cross-sectional area is proportional to the electric field strength.

Note that electric fields exist in three dimensions, that is, throughout a region of space, but it is much easier to show the patterns in two dimensions (just as with patterns of magnetic fields).

# E Experiment FF2
## Electric field patterns

The aim of this experiment is (a) to investigate the shape of several electric fields, produced by different arrangements of electrodes, and (b) to find out how the electric field between two plates depends upon the p.d. between the plates and their separation.

### Q 2.6 Study question

(a) What is the convention that is adopted for the direction of an electric field?
(b) Sketch the electric field pattern for each of the following situations (i) an isolated positive charge, (ii) an isolated negative charge, (iii) between two oppositely charged horizontal parallel plates. ■

## Uniform electric field

The results of experiment FF2(b) can be summarised as follows:

(i) The electric field between the plates is uniform.
(ii) Increasing the p.d. or decreasing the separation increases the electric field strength.
(iii) When the p.d. and the separation are both doubled the electric field strength remains the same.

We will now develop an argument which will show how the electric field strength depends upon the p.d. and the spacing.

Consider a positive charge $Q$ at a point in a uniform field between two charged plates (figure 2.3).

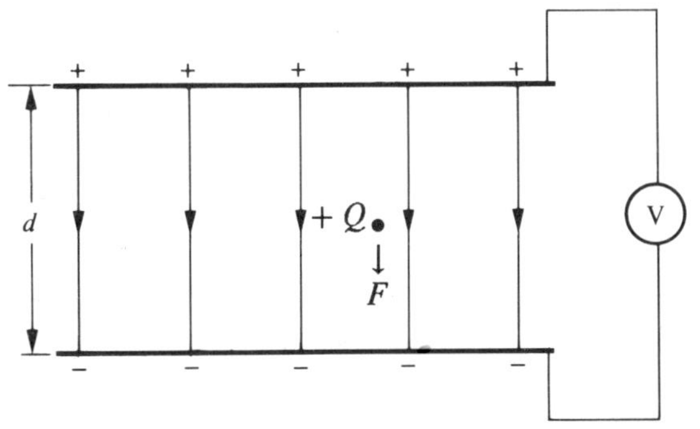


**Figure 2.3**

The positive charge will experience a force $F$ in the direction of the field. To move this charge in opposite direction to the field, work must be done by an external agent. The work $W$ depends upon the magnitude of the force and the distance moved.

This is rather like lifting a load of mass $m$ through a vertical height $h$, when the work done is equal to $mgh$ (assuming that $g$ is constant over this distance).

### Q 2.7 Development question *

(a) Show that the work done $W$ when the charge $Q$ is moved between the two plates (figure 2.3), which are separated by a distance $d$, is given by

$$W = EQd$$

(b) A voltmeter connected between the two plates gives a reading $V$ when there is an electric field between the plates. The voltmeter measures the potential difference between the plates, that is, it tells us how much work is done in moving unit charge between the two plates. Write down an equation giving the work done $W$ when a charge $Q$ is moved between two points having a p.d. of $V$.

(c) By combining the results in (a) and (b) show that the electric field strength $E$ between two plates with a p.d. of $V$ and separated by a distance $d$ is given by

$$E = \frac{V}{d} \quad ■$$

This is a very convenient relationship for a uniform electric field because it enables us to find the electric field strength by using a voltmeter and a ruler and thus avoids the rather difficult task of measuring forces on charges.

The quantity $V/d$ is known as the potential gradient and the unit in which it is measured is the volt per metre, V m$^{-1}$. This is a more usual unit for electric field strength than the newton per coulomb: the two are, of course, equivalent.

*Note.* Later in the chapter we will derive a more general relationship between electric field strength and potential gradient.

## Gravitational field

We have seen that the concept of a field of force has been introduced to describe the interaction between two charged bodies. Similarly, it can be used to describe the gravitational forces between two particles or bodies (figure 2.4).

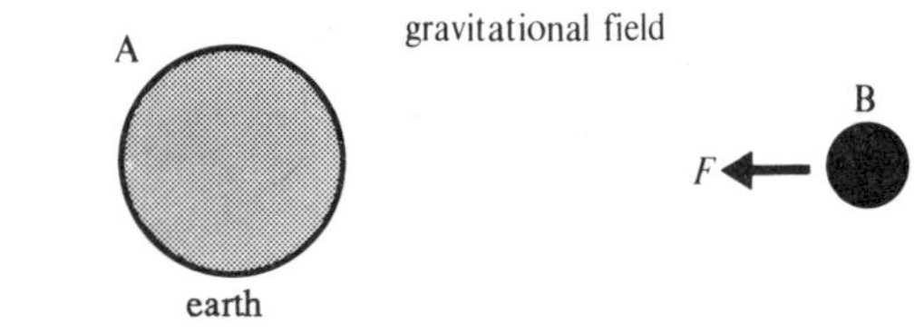


**Figure 2.4**

We suppose that

(i) a body A, such as the earth, produces a gravitational field because of its mass,
(ii) a body B, when placed in the field, will experience a force depending on the mass of B.

## Gravitational field strength

The strength of the gravitational field, $g$, can be defined in a similar way to electric field strength, by considering the force that would be experienced by a small mass if it were placed in the field.

### Q 2.8 Study question

Write a definition of the gravitational field strength $g$. Give the equation relating $g$ to a small mass $m$ and the gravitational force $F$ on it; hence write down a unit in which $g$ can be measured. Is $g$ a vector or a scalar quantity? ■

Measurements show that the earth's gravitational field strength is approximately 9.8 N $kg^{-1}$. However, when a body falls freely under gravity its acceleration is 9.8 m $s^{-2}$. Thus, we have two ways of looking at $g$.

### Q 2.9 Self-assessment question

Why are the (numerical) values of the gravitational field strength and the acceleration under gravity the same? ■

### Q 2.10 Self-assessment question

In her spacesuit, an astronaut weighs 800 N on the earth, where the gravitational field strength is 10 N kg $s^{-1}$. What would she weigh

(a) on a smaller planet, where the field strength is 2.0 N $kg^{-1}$,
(b) on a larger planet, where the field strength is 15 N $kg^{-1}$? ■

## 2.2 Radial field

### Electric field strength and distance

In the first topic we saw that the electric force between two charges depended upon the distance between them.

We now wish to find how the electric field strength is related to the distance from a point charge.

Suppose we consider a small positive charge $Q_0$ at a distance $r$ from a positive point charge $Q$ (figure 2.5).

From Coulomb's law the force exerted on this charge is

$$F = \frac{1}{4\pi\varepsilon_0} \cdot \frac{Q\,Q_0}{r^2}$$

### Q 2.11 Development question

(a) The electric field strength $E$ at a given point in the field is equal to the force per unit charge. Hence show that

$$E = \frac{1}{4\pi\varepsilon_0} \cdot \frac{Q}{r^2}$$

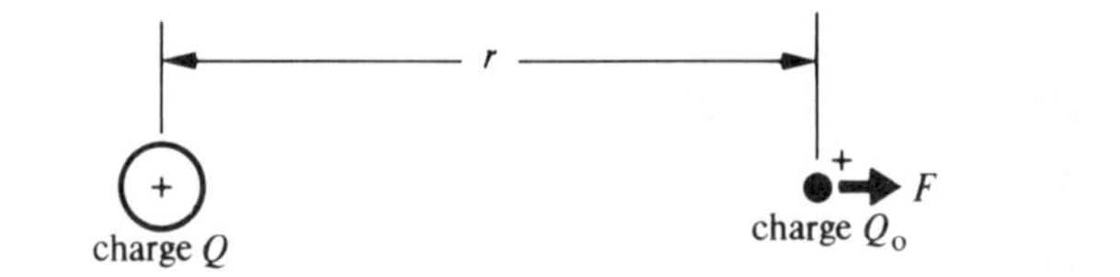


Figure 2.5

(b) From consideration of symmetry, what can you say about the direction of the field (i) when $Q$ is positive and (ii) when $Q$ is negative? ■

The **intensity** of an **electric field** due to a point charge varies **inversely** as the **square** of the **distance** from the point (figure 2.6).

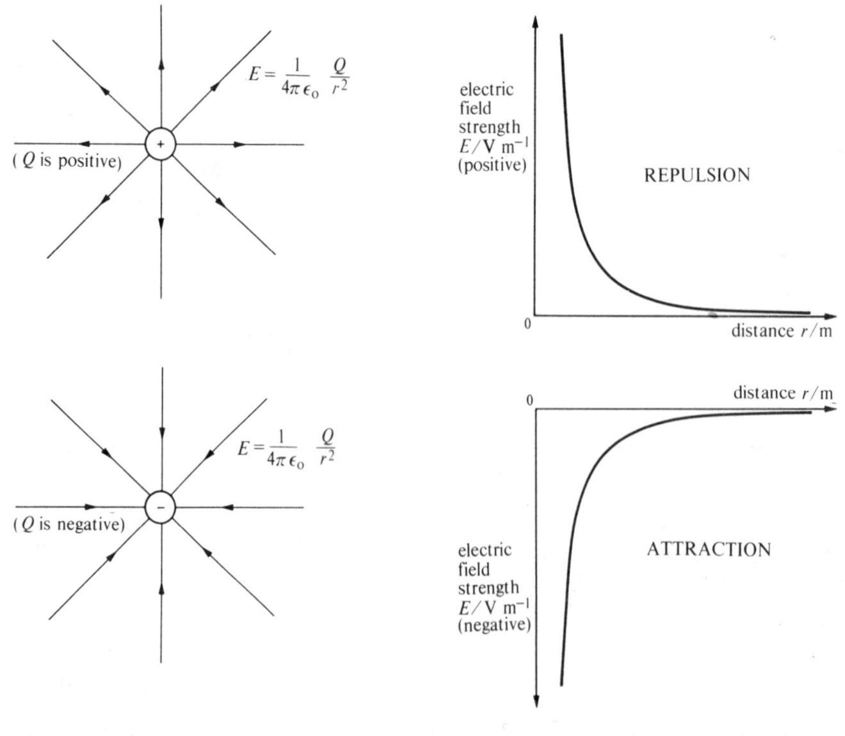


Figure 2.6

## Q 2.12 Self-assessment question

(a) Why does repulsion correspond to a positive value for $E$ and attraction to a negative value?
(b) Two point charges A and B, magnitudes $1.0 \times 10^{-9}$ C and $2.0 \times 10^{-9}$ C respectively, are placed 0.9 m apart (in a vacuum or air), and there are no other charges in the vicinity. What is the magnitude and direction of the electric field strength at a point P which lies on the line joining A and B and is 0.3 m from A
(i) if both charges are positive,
(ii) if both charges are negative,
(iii) if the charge at A is positive and the one at B negative,
(iv) if the charge at A is negative and the one at B positive?■

In the unit *Behaviour of matter* there is an account of the way in which electric attractive forces between particles in a molecule, for example sodium chloride, give the solid its rigid structure. In sodium chloride the sodium atom has transferred an electron to the chlorine atom, and from our knowledge of Coulomb's law it is possible to calculate, to an order of magnitude, the attractive force between the neighbouring sodium and chloride ions.

## Q 2.13 Self-assessment question

Suppose we assume that the separation of the sodium ion and the chloride ion is $2.0 \times 10^{-10}$ (which is approximately one atomic diameter) and that the magnitude of the charge on each ion is $1.6 \times 10^{-19}$ C.

(a) Calculate the electric field strength at a distance of $2.0 \times 10^{-10}$ m from a charge of $1.6 \times 10^{-19}$ C, in a vacuum.
(b) What force is exerted on a charge of $1.6 \times 10^{-19}$ C in a field of this magnitude? ■

## Gravitational field strength and distance

Consider a body of small mass $m$ at a distance $r$ from the centre of the earth, mass $M$ (figure 2.7)

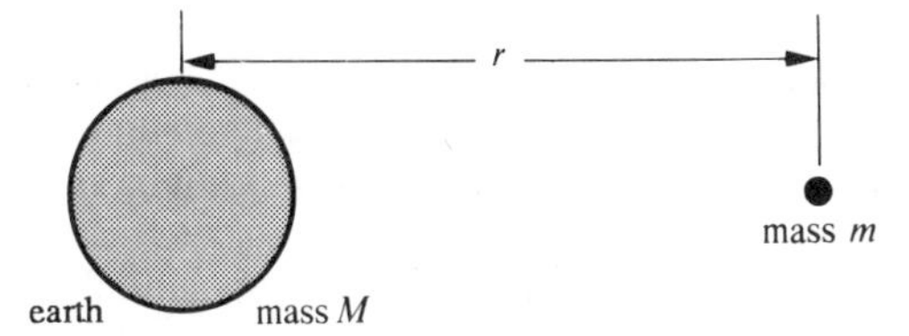


**Figure 2.7**

*Note.* We assume that the earth behaves as if its whole mass was concentrated at its centre. This can be shown to be correct by mathematics (which is beyond the requirements of this course) if the mass is symmetrically placed round the centre, in uniform layers, as is approximately the case for the earth.

It is clear that the field around any spherically symmetrical distribution of mass must also be spherically symmetrical; the only question is about its magnitude.

From Newton's law of gravitation, the force exerted on the mass $m$ is

$$F = \frac{GMm}{r^2}$$

*Note.* This equation tells us the magnitude of the force. The direction of the force $F$ is parallel but in the opposite direction to the distance $r$.

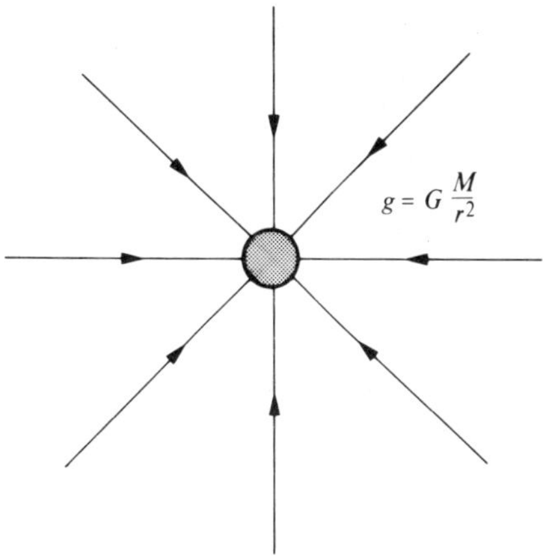


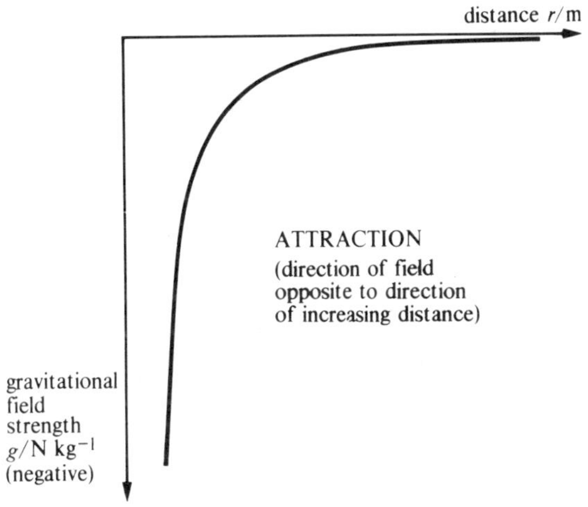


**Figure 2.8**

## Q 2.14 Development question

The strength of the gravitational field $g$ at a point in the field is equal to the force on the unit mass. Hence, show that

$$g = \frac{GM}{r^2}$$

*Note.* This is also a relationship betwen $g$ and $G$. The direction of the field is radially inwards. ■

The **gravitational field strength** due to the earth varies outside the earth **inversely** as the **square** of the **distance** from the centre of the earth (figure 2.8).

**Q 2.15 Self-assessment question**

The gravitational field strength at the surface of a planet of radius $R$ is $g_p$. What is the gravitational field strength at a distance from the centre of (a) $2R$ and (b) $100R$? ■

If something is flowing steadily and uniformly out from the centre of a sphere, the further it gets from the centre the larger the sphere over which it is spread. Because the surface area of a sphere is proportional to $r^2$, the surface density (the amount per unit area) of whatever is spreading out is inversely proportional to $r^2$. The inverse square law can be thought of as a natural consequence of our three-dimensional world applying, for example, to light. Both electric and gravitational fields can be thought of in terms of something (called flux) imagined to flow out from the centre. The lines of field are very similar to lines of this flux.

**Q 2.16 Self-assessment question**

(a) Explain why there will be a point between the earth and the moon where a spaceship will experience zero resultant force due to the earth-moon system.
(b) Show that, at this point,

$$\left(\frac{r_1}{r_2}\right)^2 = \frac{M_E}{M_M}$$

where $r_1$ and $r_2$ are the distances from the earth and the moon respectively to the point, and $M_E$ and $M_M$ are the masses of the earth and moon respectively.
(c) Calculate a value for the gravitational field strength at the surface of the moon, given that the radius of the moon is $1.7 \times 10^6$ m and the average density of its material is $3.4 \times 10^3$ kg m$^{-3}$.

State clearly the assumption that you have made.
($G = 6.7 \times 10^{-11}$ N m$^2$ kg$^{-2}$) ■

**Q 2.17 Self-assessment question**

(a) Suppose a stellar object were to be discovered with negative mass, perhaps a whole galaxy. What would the gravitational field round it be like?
(b) What would the orbit of a planet of negative mass be near a star of negative mass?
(c) What would the orbit of a planet of negative mass be near an ordinary star of positive mass? ■

EXTENSION

**Q 2.18 Study question**

Make notes including the derivation of relevant formulae on the following topics

(a) variation of $g$ with height above the earth's surface,
(b) variation of $g$ with depth, for example in a mine. (State the assumption that you have made.) ■

## 2.3 Potential energy in a uniform field

### Potential difference

Scientists and technologists use the word **work** in a more precise way than that used in everyday speech.

**Q 2.19 Self-assessment question**

Give this precise meaning of work. ■

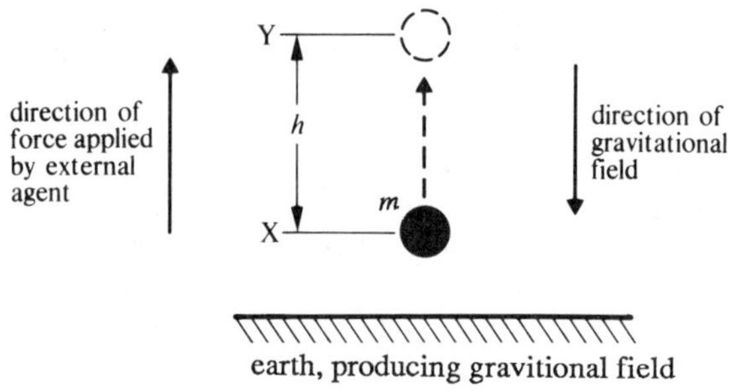


**Figure 2.9**

**Q 2.20 Self-assessment question**

(a) Assuming that the earth's gravitational field strength $g$ is constant (figure 2.9) show that the work done by the external agent, in lifting a mass $m$ through a vertical height $h$, is given by

$$W = mgh$$

(b) What energy transfer is measured if you lift a stone in your hand? ■

We say that the system has gained and stores in the gravitational field this amount of **gravitational potential energy** (g.p.e.).

**Q 2.21 Self-assessment question**

(a) Calculate the work done by an external agent in lifting an object of mass 5.0 kg through a vertical height of 2.0 m above the earth's surface in a uniform gravitational field of 10 N kg$^{-1}$.
(b) What is the increase in gravitational potential energy of the object?
(c) How much work has been done on the field? ■

On this basis we can look at fields in a new way.

The gravitational potential energy associated with the object depends upon its location in the field of force and the magnitude of the mass. Thus, if we state the magnitude of the mass we can describe the gravitational field in terms of the potential energy of the mass at different points in the field. Unit mass is chosen and the change in potential energy per unit mass which occurs when a small mass is moved from one point to another is called the change in **gravitational potential**, or the **gravitational potential difference** $\Delta V_g$ between the two points in the field.

Thus, we can write

work done = mass x gravitational potential difference

$$W = m\ \Delta V_g$$

## Q 2.22 Self-assessment question

What is the gravitational potential difference between two points which are situated 2.0 m apart along the same field in a uniform gravitational field?
(gravitational field strength $g$ = 10 N kg$^{-1}$) ■

Each point in a gravitational field can be described in terms of **potential**. This is numerically equal to the work that is done per unit mass in bringing a small mass to that point from a place where the potential is zero. The choice of **zero potential** is arbitrary: sometimes the surface of the earth is taken as the zero, in other cases zero potential is taken as the potential at an infinite distance from the earth. (Later in the topic we will consider the potential at a point in an inverse square law gravitational field.)

Similarly, each point in an electric field has an **electric potential.**

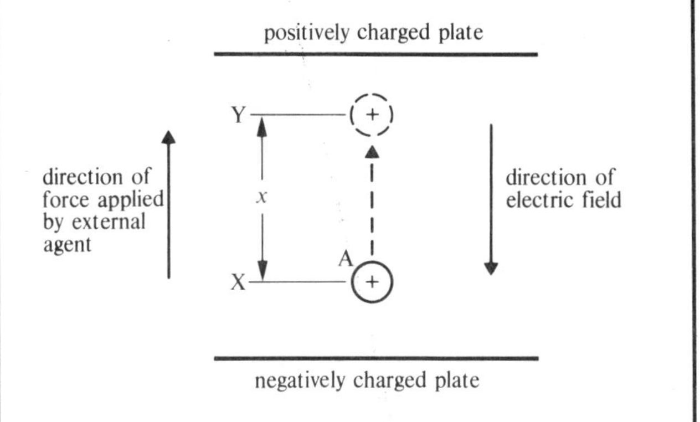


**Figure 2.10**

Consider a positive charge of magnitude $Q_0$ at a position X in an electric field (figure 2.10). The positive charge will experience a force $F$ in the direction of the field. To move this charge from X to Y work must be done by an external force and energy has to be supplied.

## Q 2.23 Self-assessment question

(a) What are the factors that the work done depends upon?
(b) Show that the work done is given by the expression

$$W = E\,Q_0 x$$

assuming the electric field strength $E$ is constant. ■

We say that the system has gained energy and stores in the electric field this amount of electric potential energy.

## Q 2.24 Self-assessment question

(a) Calculate the work done in moving a positive charge of 2.0 μC between two points 0.1 m apart in an electric field of strength 5.0 N C$^{-1}$.
(b) What is the increase in electric potential energy?
(c) How much work is done when a charge of one coulomb is moved between the two points? ■

## Q 2.25 Study question

(a) Making the comparison with the gravitational case, explain what is meant by the electric potential difference between two points in an electric field, and give the unit on this basis (equal to the volt).
(b) Write an equation for the work done in terms of charge and potential difference.
(c) What conventions are used in theoretical work, and for practical purposes, about the zero of electric potential? ■

Later in the topic we will consider the potential at a point in an inverse square law electric field.

## Q 2.26 Self-assessment question

(a) The work that is done by an external agent in bringing a positive charge of 0.2 μC from a place of zero potential to a point P in an electric field is 6.0 x 10$^{-4}$ J. Calculate the potential at the point P.
(b) If 4.0 x 10$^{-4}$ J of work is done by the electric field when the charge moves to a point Q, what is the potential difference between P and Q? What is the potential at the point Q?
(c) How much work must be done in bringing a charge of 0.3 μC from a place of zero potential to the point Q? ■

# Equipotentials

All points in a field which are at the same potential may be imagined to be connected together; the surface which is formed is known as an equipotential surface (or a line in a two-dimensional diagram).

## Q 2.27 Study question

Draw diagrams to show how the electric field can be represented by equipotential lines in each of the following cases

(a) the radial field of a point positive charge,
(b) the uniform field between two charged plates. ■

## Q 2.28 Self-assessment question

How much work is done in moving a test charge between any two points on an equipotential surface? ■

## Q 2.29 Self-assessment question

What is the shape of a gravitational equipotential surface close to the earth's surface? ■

## Q 2.30 Study question*

Make a brief summary of the two ways in which we can describe either an electric field or gravitational field. ■

# Potential gradient

We will now consider how the potential is related to the electric field strength.

## Q 2.31 Development question

Consider a small positive charge $Q_o$ at a point A in an electric field where the electric field strength is $E$ (figure 2.11). The direction of the electric field is defined to be the positive direction.

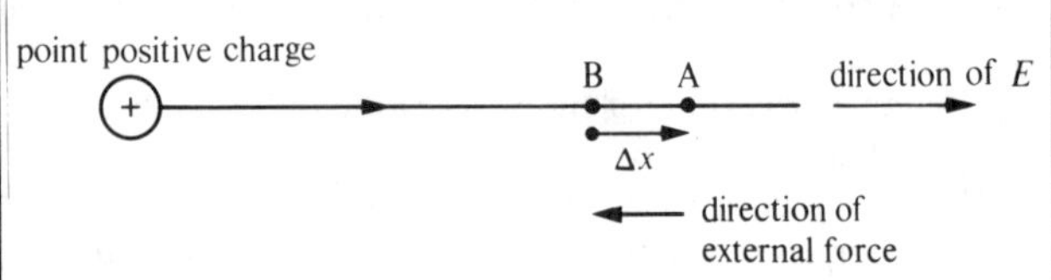


**Figure 2.11**

(a) What is the magnitude and direction of the electric force which is exerted on the test charge at A?

Suppose that the charge is pushed by an external force a very short distance $\Delta x$ to B (in the opposite direction to $E$).

(b) Show that the work $\Delta W$ that has to be done on the charge (by this external agent) in moving from A to B (assuming $E$ remains constant over this distance) is given by

$$\Delta W = (-EQ_o)\ \Delta x$$

*Note.* (i) The negative sign arises because the force is in the negative $x$ direction. (ii) In the move from A to B,Δ ξ is negative and so $\Delta W$ is positive (showing that work is done on the charge).

(c) Is the potential at B greater or less than at A?

Suppose that the p.d. between A and B is $(V_B - V_A) = \Delta V$. This is positive, indicating an increase in potential from A to B.

(d) Remembering the definition of p.d. show that

$$E = -\frac{\Delta V}{\Delta x}$$ ■

In the limit as $\Delta x$ tends to zero we have

$$E = -\frac{dV}{dx}$$

This equation is quite general, although we have derived it in one particular situation.

The quantity $dV/dx$ is called the **potential gradient** (the rate of change of potential with displacement).

*Note.* (i) The negative sign tells us that the electric field strength is in the opposite direction from that of the potential gradient. (ii) In the case of a uniform electric field, e.g. between two parallel plates

$$E = -\frac{V}{d}$$

where $V$ is the p.d. between the plates and $d$ is the separation.

## Q 2.32 Self-assessment question

Show that the volt per metre V m$^{-1}$ is equivalent to the newton per coulomb N C$^{-1}$ as a unit for electric field strength. ■

## Q 2.33 Self-assessment question

A small charged strip of foil (highly flexible) on an insulating handle is held between two large vertical plates which are connected to a 2500 V supply. The distance between the plates is 50 mm. The deflection of the strip is noted.

(a) The plates are moved until they are 100 mm apart (the p.d. is kept constant). Will the strip hang at a larger or smaller angle? Give a reason.

(b) Calculate the strength of the electric field in the two situations.

(c) To what value must the p.d. be changed in order that the force exerted on the strip is the same as that at the smaller spacing? ■

## Q 2.34 Self-assessment question

This question illustrates the principle of a method of measuring the elementary charge.

Two horizontal parallel plates are 20 mm apart. An electric field is maintained between the plates by a potential difference of $10^4$ V. An electrically charged oil drop of mass $10^{-7}$ kg is seen to be suspended motionless between the plates.

(a) Draw a diagram to show the forces acting on the oil drop.

(b) Calculate the charge on the oil drop.

(c) What additional potential difference must be applied to produce a vertical acceleration in the electric field of 0.02 m s$^{-2}$ upwards? (Ignore air resistance and Archimedian upthrust.) ■

# The electronvolt

A non-SI unit which is commonly used when dealing with electrons and ions is the electronvolt.

When a charge $Q$ moves through a potential difference of V, the energy $W$ acquired is given by $W = QV$.

If the charge is equal to the charge on the electron (1.6 x $10^{-19}$ C), and the potential difference is 1 V, then the energy change is given by

$$W = 1.6 \times 10^{-19}\ \text{C} \times 1.0\ \text{V} = 1.6 \times 10^{-19}\ \text{J}$$

This quantity of energy is called the electronvolt, eV. It is the energy required by an electron when its moves through a potential difference of 1 volt.

## Q 2.35 Self-assessment question

An electron in an X-ray tube is accelerated from rest through a potential difference of 18 000 V. Calculate, for the moment when the electron hits the target of the X-ray tube,

(a) its kinetic energy in electronvolts,

(b) the speed of the electron.

(Kinetic energy = $mv^2/2$. The magnitude of the ratio of charge to mass for an electron, $e/m_e$ = 1.8 x $10^{11}$ C kg$^{-1}$) ■

## Q 2.36 Study question

Show that for a gravitational field, the gravitational field strength is equal to the negative potential gradient, that is

$$g = -\frac{dV}{dx}$$

*Note.* There is always a negative sign in the equation, field $= -dV/dx$. This is because movement in the same direction as the field results in a decrease in potential. The distance $x$ is taken to increase along the field direction. ■

For a uniform gravitational field, for example near the surface of the Earth, we can write

$$g = -\frac{\Delta V_g}{\Delta x}$$

where $\Delta V_g$ is the gravitational potential difference between two points in the field which are separated by $x$.

**Table 2.1 Comparison of electric and gravitational fields**

| Electric field | Gravitational field |
|---|---|
| electric field strength | gravitational field strength |
| $E = F/Q$ | $g = F/m$ |
| electric potential difference (uniform field) | gravitational potential difference (uniform field) |
| $\Delta V_E = W/Q$ | $\Delta V_g = W/m$ |
| electric potential gradient | gravitational potential gradient |
| $E = -dV_E/dx$ | $g = -dV_g/dx$ |

## 2.4 Potential in a radial field

In the first section of this topic we saw that, in both electric and gravitational fields, the field intensity at a point near a spherical source (charge or mass) is inversely proportional to the square of the distance from the centre of the source. We now want to investigate how the potential at a point in such a field depends upon distance. We will first consider an electric field.

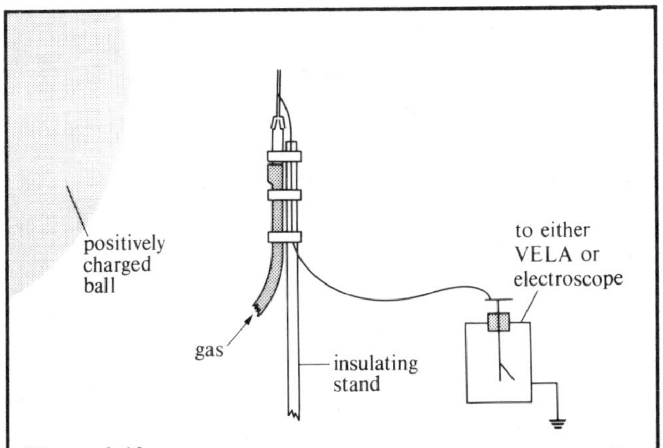


**Figure 2.12**

Just as **gravitational potential difference** is the change in energy **per unit mass** in moving between two points, **electric potential difference** is the change in energy **per unit charge** in moving between two points. The unit of energy per unit charge, i.e. the joule per coulomb is, by definition, the volt and hence p.d. is measured in volts.

It is not possible to use a moving-coil voltmeter because this type of instrument requires the continuous flow of charge for its operation. Hence we use a flame probe which consists of a hypodermic needle with a small gas flame at the end.

The probe (Figure 2.12) is connected to a high impedance voltmeter or an electroscope which has been calibrated as a voltmeter. The flame ionises the air around the tip. Charges accumulate until the tip acquires the same potential as the field locally. The voltmeter is then capable of measuring the potential at the tip.

It is important to realise that the probe measures the difference in potential between the tip and the potential at infinity. *This is defined arbitrarily to be zero.*

### Q 2.37 Development question*

By answering the following question you should be able to explain the action of the flame probe, that is, how an ion source senses the potential at a point in an electric field.

Suppose that the probe, without the flame, was brought up to a place near the positively charged sphere.

(a) What type of charge would appear on the probe?
(b) What type of charge would appear on the voltmeter?

Now suppose that the flame is lit.

(c) What is produced in the vicinity of the probe?
(d) Why will the tip of the probe eventually end up with no charge? ■

The voltmeter remains positively charged and since it is connected to the probe it shows the potential at the position of the probe.

---

### E Experiment FF3
**The potential around a sphere**

In this experiment you will use a flame probe to investigate how the potential in the electric field of a large charged sphere varies with distance.

---

The results of this experiment show that the potential $V$ at a point in the electric field due to a charged sphere is inversely proportional to the distance $r$ from the point to the centre of the sphere. That is

$$V \propto \frac{1}{r}$$

### Q 2.38 Self-assessment question

A sphere of radius 0.1 m is charged to a potential of 6.0 kV.

In the absence of any other charged bodies what will be the potential at a point (a) 0.2 m, (b) 0.3 m and (c) 0.4 m from the centre of the charged sphere? ■

## Potential at a point in an electric field

We will now derive an expression for the potential at a point in an electric field.

### Q 2.39 Development question*

Consider a point P in the radial field produced by a point positive charge $+Q$ situated at O in a medium of permittivity $\varepsilon_0$ (a vacuum or air), figure 2.13.

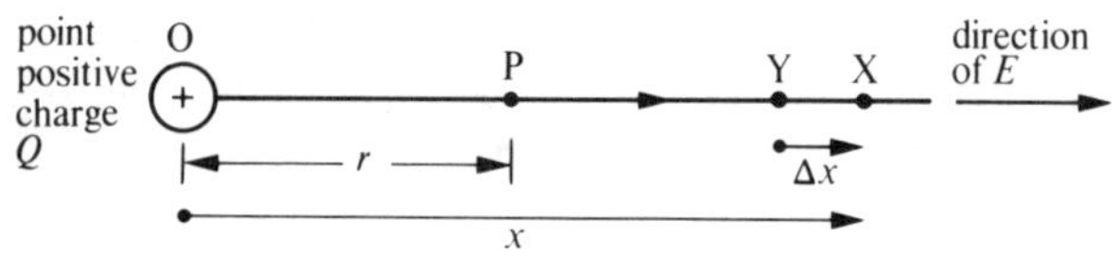


**Figure 2.13**

To find the potential at P we have to calculate the work done in bringing a unit positive charge from infinity (where the potential is assumed to be zero) to the point P. Because the field is not constant over this distance (it varies as $1/r^2$) we must consider the movement of the charge over a large number of small steps and find the sum of the work done in each step.

Imagine that a small positive charge $Q_0$ has been brought from infinity and is at a point X a distance $x$ from O.

(a) Write down an expression for the electric force $E$ which is exerted on this charge.
Suppose that the charge is moved by an external agent a very short distance $\Delta x$ to Y.
(b) Show that the work $W$ that has to be done on the charge in moving it from X to Y (assuming that $E$ remains constant over this distance) is given by

$$W = \frac{1}{4\pi\varepsilon_0} \cdot \frac{QQ_0}{x^2} \Delta x$$

*Note*. The negative sign arises because the force is in the negative $x$ direction.

Figure 2.14 shows how the force which is exerted on the charge by the external agent varies as the charge is brought closer and closer to 0.

(c) What does the area of the narrow strip represent?
(d) State in words how you would find the work done $W$ by the external agent in bringing the charge from infinity to a distance $r$ from O. ■

It can be shown by integral calculus that the work done is given by

$$W = \frac{1}{4\pi\varepsilon_0} \cdot \frac{QQ_0}{r}$$

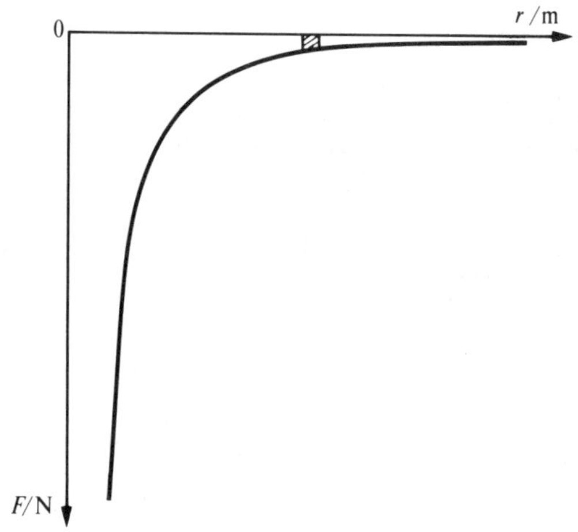


**Figure 2.14**

Since the potential at a point is the work done per unit of charge we have

$$V = \frac{W}{Q_0} = \frac{1}{4\pi\varepsilon_0} \cdot \frac{Q}{r}$$

The **electric potential** at a point in an electric field due to a point charge is **inversely proportional** to the **distance** from the point charge (figure 2.15). Note the differences between the graphs for field and for potential.

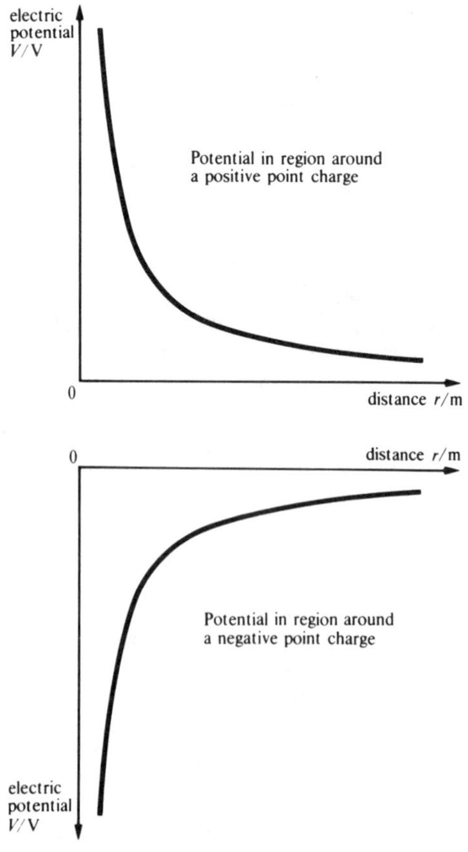


**Figure 2.15**

**EXTENSION**

We can calculate the work done by integration:

$$W = -\int_{\infty}^{r} \frac{1}{4\pi\varepsilon_0} \cdot \frac{QQ_0}{x^2}\,dx = \frac{-QQ_0}{4\pi\varepsilon_0}\int_{\infty}^{r}\frac{dx}{x^2}$$

$$W = \frac{1}{4\pi\varepsilon_0}\left[\frac{QQ_0}{x}\right]_{\infty}^{r}$$

$$W = \frac{1}{4\pi\varepsilon_0}\frac{QQ_0}{r}$$

---

## E Experiment FF4
**Relationship between charge and potential**

In this experiment you will measure charge and determine a value for $1/4\pi\varepsilon_o$.

---

The numerical value you should have obtained for $1/4\pi\varepsilon_o$ from this experiment should be somewhere close to 9 x $10^9$.

### Q 2.40 Self-assessment question

(a) Show that the unit of $1/4\pi\varepsilon_o$ is N $m^2$ $C^{-2}$.
(b) What is the value for $\varepsilon_o$ ? ■

### Q 2.41 Self-assessment question

What is the potential at a point 2.0 m from a point positive charge of 1.0 μC in a vacuum?

$$\frac{1}{4\pi\varepsilon_0} = 9.0 \times 10^9 \text{ N m}^2 \text{ C}^{-2}$$ ■

### Q 2.42 Self-assessment question

In this question you will make an estimate of the energy that is required to ionise a hydrogen atom. A hydrogen atom consists of a proton and an electron at an average separation of 0.5 x $10^{-10}$ m.

(a) What is the electric potential at 0.5 x $10^{-10}$ m from the proton?
(b) What is the electric potential energy of the electron-proton system?
(c) How much energy would be required to remove the electron from the atom if we imagine that the electron is at rest?

If we assume that the electron is in motion around the proton it will also have kinetic energy. Consider the electron moving in a circle of radius *r* with a speed *v*.

(d) Write down an expression for the force which provides the centripetal force.
(e) Hence show that the kinetic energy is given by

$$E_k = \frac{1}{2}mv^2 = \frac{1}{2}\cdot\frac{1}{4\pi\varepsilon_0}\cdot\frac{e^2}{r}$$

where *e* is the magnitude of the charge on the proton and the electron.

(f) What is the kinetic energy of the electron?
(g) What is the total energy of the electron-proton system? ■

## Gravitational potential

In order to gain an understanding of the way in which gravitational potential varies in the earth's gravitational field we will solve problems using some more information from the Apollo space flight (figure 2.16).

**Figure 2.16**

The information in table 2.2 covers two periods in the flight, one on the outward journey to the moon and one on the return to earth. Within both periods the motors were not in use, though they were used between these periods to land on the moon and take off again.

**Table 2.2**

| Distance from centre of earth $r/10^6$ m | Speed $v$/m $s^{-1}$ | Kinetic energy per kilogram $\frac{1}{2}v^2/10^6$ J $kg^{-1}$ |
|---|---|---|
| **Outward** | | |
| (1) 11.0 | 8406 | 35.33 |
| (2) 26.3 | 5374 | 14.44 |

continued

| $r/10^6$ m | $v$/m $s^{-1}$ | $\frac{1}{2}v^2/10^6$ J $kg^{-1}$ |
|---|---|---|
| (3) 54.4 | 3653 | 6.35 |
| (4) 95.7 | 2619 | 3.43 |
| (5) 169.9 | 1796 | 1.61 |
| (6) 209.2 | 1532 | 1.17 |
| (8) 240.6 | 1356 | 0.92 |
| **Return** | | |
| (9) 241.6 | 1521 | 1.16 |
| (10) 209.7 | 1676 | 1.40 |
| (11) 170.9 | 1915 | 1.83 |
| (12) 96.8 | 2690 | 3.62 |
| (13) 56.4 | 3626 | 6.57 |
| (14) 28.4 | 5201 | 13. 52 |
| (15) 13.3 | 7673 | 29. 44 |
| (16) 10.0 | 8854 | 39. 20 |

## Q 2.43 Development question*

(a) What is the change in the kinetic energy of each kilogram of the spacecraft between points 1 and 6 on the outward journey? Account for this change in energy.
(b) What is the change in the kinetic energy of each kilogram of the spacecraft between points 10 and 16 on the return journey? Account for it.
(c) How do you think your answers to (a) and (b) would have compared if the pairs of distances had been equal?
(d) What name is given to the change in gravitational potential energy per kilogram between two points in a gravitational field?
(e) What is the gravitational potential difference between points 1 and 2? ■

It is also possible to calculate the gravitational potential difference from our knowledge of the way in which the force varies between the two points.

## Q 2.44 Development question*

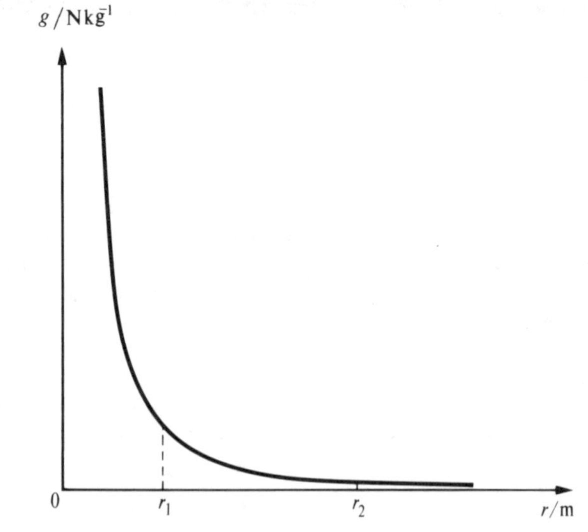


**Figure 2.17**

Consider a point a distance $r$ from the centre of the earth.

(a) Write down an expression for the gravitational force which is acting on the spacecraft.
Suppose the spacecraft moves a distance $\Delta r$ from the centre of the earth.
(b) Write down an expression for the gain in potential energy per kilogram.
Figure 2.17 shows how the gravitational field strength, i.e. the force per kilogram varies with distance.
(c) State in words how you would find the gain in potential energy per kilogram as the spacecraft moves from $r_1$ to $r_2$.■

The area of each strip is the change in potential that occurs between the two limits. The total change in potential can be found by adding up all the strips.
The result of this shows that the gravitational potential difference ($\Delta V_g$) between two points is

$$\Delta V_g = GM\left(\frac{1}{r} - \frac{1}{r^2}\right)$$

*Note*: At this stage we have not given the minus signs appropriate to the convention that the gravitational potential is zero at large distances.

**EXTENSION**

This area can also be found by integration as follows:

$$\text{gain in potential energy per kilogram} = \int_{r_1}^{r_2} \frac{GM}{r^2} \cdot dr$$

$$= -\left[\frac{GM}{r}\right]_{r_1}^{r_2}$$

$$= \frac{-GM}{r_2} - \left(\frac{-GM}{r_1}\right)$$

$$= \frac{GM}{r_1} - \frac{GM}{r_2}$$

## C Computer program GFIELD

This program calculates the gravitational potential difference between the points $r_2$ and $r_1$ in figure 2.17 by splitting the area into small blocks. The field is assumed to be constant within each block. You should follow the exercises in the notes accompanying the program.

## Q 2.45 Development question*

For the earth $GM = 4.0 \times 10^{14}$ N $m^2$ $kg^{-1}$ and for points 1 and 2, $r_1 = 11.0 \times 10^6$ m and $r_2 = 26.3$ m. Use these values to calculate a value for the gain in potential energy per kilogram, that is, the gravitational potential difference between the two points. How does it compare with your value from the spaceflight data? ■

## Q 2.46 Development question*

In this question you will use the information in table 2.2 to show the variation of potential with distance.

(a) At $r = 11.0 \times 10^6$ m on the outward journey, the kinetic energy of each kilogram of the spacecraft is $35.33 \times 10^6$ J kg$^{-1}$. At $r = 240.6 \times 10^6$ m it is only $0.92 \times 10^6$ J kg$^{-1}$. By how much has the potential energy per kilogram changed?
(b) Is it possible that at some distance $r_0$ the kinetic energy would become zero?
(c) What would be the change in potential energy per kilogram between $r = 11.0 \times 10^6$ and $r_0$?
(d) What would be the change in potential energy per kilogram between $r = 26.3 \times 10^6$ m and $r_0$?
(e) Write an equation for the change in potential energy per kilogram between $r$ and $r_0$.
(f) The increase in potential energy per kilogram between distance $r$ and distance $r_0$ is equal to the decrease in kinetic energy per kilogram given against $r$ in the table. Plot graphs of the magnitude of the potential energy between distance $r$ and the (unknown) distance for both the outward and homeward sets of data, against $1/r$. $r_0$ is not the same for these two sets of data, but is constant for all the data in one set.
(g) What should the shape of each of the two graphs be, if the equation given in the answer to (e) is correct?
(h) What should the slopes of the graphs be? Should they have the same slope?
(i) What intercept should each graph have with the axis of $1/r$? Will this be the same for both graphs? ■

### EXTENSION

### Potential at a point in a gravitational field

To find the potential at a point P a distance $r$ from a point mass $M$ at O we have to find the work done in bringing unit mass from infinity to the point P, figure 2.18. (Our zero of gravitational potential is chosen (by convention) to be a point as far away as possible from all other matter, that is at infinity.)

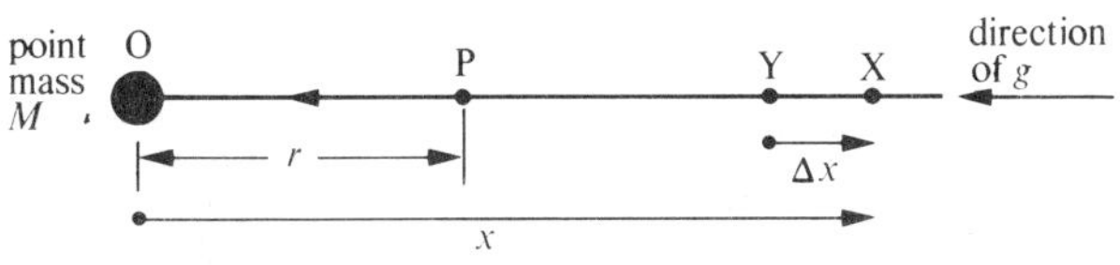


**Figure 2.18**

## Q 2.47 Study question*

(a) Show that the potential at a point in a gravitational field is given by

$$V = -\frac{GM}{r}$$

remembering that the positive direction is away from the point mass.

continued

(b) What is the significance of the negative sign in the equation? ■

The **gravitational potential** at a point in a gravitational field is **inversely proportional to the distance** (figure 2.19). Note the differences between the graphs for field and for potential.

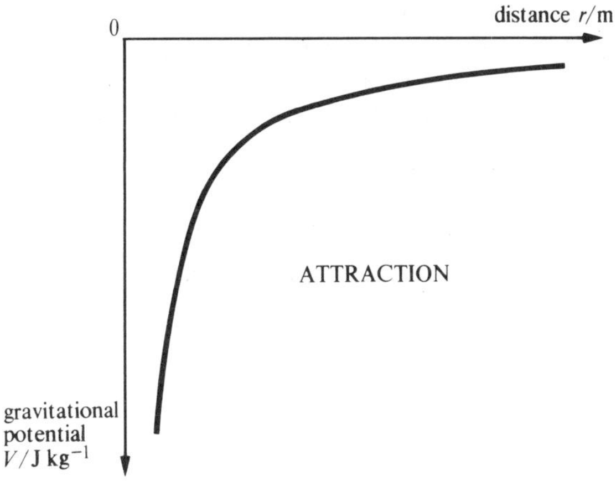


**Figure 2.19**

## Q 2.48 Self-assessment question

(a) Calculate the gravitational potential at a point on the earth's surface. The radius of the earth is $6.4 \times 10^6$ m.
(b) What is the kinetic energy per kilogram that must be possessed by a rocket if it is to escape from the earth's gravitational field. ? ■

## Q 2.49 Self-assessment question

(a) Write down an expression for the potential $V$ at a distance $r$ from a point negative charge of magnitude $Q$.
(b) Is the potential negative or positive? ■

## Q 2.50 Study question

Explain the meaning of the following
(a) a conservative field, (b) a potential well. ■

## Q 2.51 Study question

(a) How does the gravitational potential vary with distance from the centre *inside* a solid uniform sphere?
(b) What difference does it make if the sphere is (like the earth) made up of layers of various different densities? ■

## 2.5 Applications of electrostatics

In this section we will consider how the shape of a charged conducting surface affects the strength of the electric field near to its surface since this has important practical applications.

### Curvature of surface and electric field strength

The electric field strength near the surface of a conductor in air at atmospheric pressure cannot be raised to more than about $3 \times 10^6$ V $m^{-1}$ because under such a potential gradient the insulating properties break down, and the air becomes ionised and behaves like a conductor. It is important, therefore, to examine the factors which determine the electric field strength near conducting surfaces. The simplest case to consider is a sphere.

**Q 2.52 Development question**

A sphere of radius $r$ is raised to a potential $V$. Show that the electric field strength just outside its surface is given by the expression

$$E = \frac{V}{r}$$ ■

For conductors which are not spherical we can consider the surface divided into a number of small regions each of which may be considered as a portion of a sphere. Since the potential $V$ is the same all over the surface of a conductor the electric field strength at any given place will be $V/r$ where $r$ is the local radius of the surface. Thus, at a place where the value of $r$ is very small the electric field strength will be high.

### Action at points

The electric field strength near a pointed conductor may be high enough to ionise the air. This effect has several important applications.

**Q 2.53 Self-assessment question**

Calculate the approximate potential to which a point of radius 0.01 mm may be raised before discharge to the atmosphere occurs. ■

**Q 2.54 Study question**

(à) Explain what is meant by 'electric wind' and point or corona discharge.
(b) Explain how a lightning conductor works. ■

The Van de Graaff generator is an electrostatic machine which can produce a high potential difference. It has been used for accelerating atomic particles.

**Q 2.55 Study question**

(a) Make notes on the Van de Graaff generator, paying attention to its construction and an explanation of how it develops a high potential.
(b) Explain why for the highest p.d.s the entire apparatus is housed in a large container which is filled with a gas at a high pressure (of the order of 20 atmospheres). ■

**Q 2.56 Study question**

During the past thirty years, there has been a rapid expansion in the industrial applications of electrostatics. Some of the more important fields are listed in table 2.3. Make brief notes on two of these. ■

**Table 2.3**

| Area of application | Topic |
|---|---|
| Energy conversion and space | Electrostatic generators; electric propulsion of space satellites; electrostatic motors |
| data processing Communications and technology | Microphones and loud speakers; liquid crystal displays |
| Surface coatings | Electrostatic painting and powder coating; electrostatic crop spaying |
| Printing | Electrostatic imaging; non impact printing photocopiers |
| Industrial processes | Mineral separation; manufacture of abrasive papers |
| Environmental processes | Precipitation of dust, smokes; air conditioning; pollution control |
| Static dischargers | Aircraft static dischargers |
| Instrumentation | Electric field meters (solid stage and rotating types); charge density meter (petroleum industry); surface charge instruments |

**Background reading**

*Physics and the engineer*, Nuffield Advanced Science.
*Electrostatic engineering*, N.J. Felici.

EXPERIMENT FF2

# Electric fields

## (a) Electric field patterns

### Apparatus

- ❍ e.h.t. power supply 0-5 kV (or van de Graaff generator)
- ❍ electric field apparatus
- ❍ castor oil
- ❍ carbon tetrachloride
- ❍ semolina
- ❍ leads

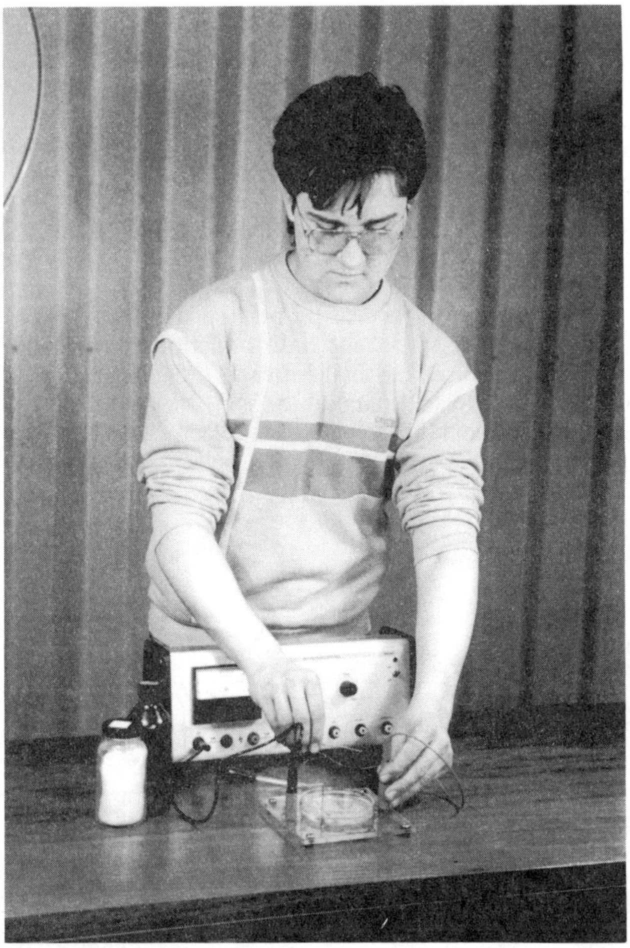

Figure 2.20

1 Arrange for the semolina to float on the interface between a layer of carbon tetrachloride and a layer of castor oil. This is done by pouring the oil into the dish, then sprinkling the semolina on top of the oil and finally adding (slowly) a layer of carbon tetrachloride.

2 Place the two straight electrodes in the petri dish and connect them to the e.h.t. power supply, as shown in figure 2.20.

3 Set the e.h.t. power supply to 4-5 kV. Observe and sketch the field pattern that is produced.

4 Repeat the experiment with the electrode arrangements shown in figure 2.21, sketching the pattern that is produced in each case:

(a) two point electrodes,
(b) a point electrode inside a ring,
(c) a point electrode and a straight electrode.

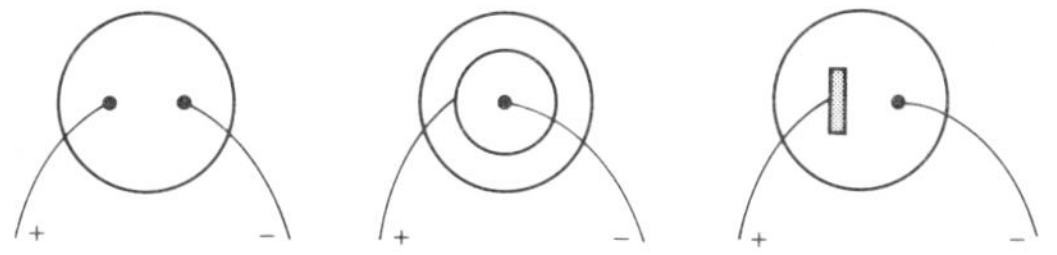


Figure 2.21

continued

**EXPERIMENT FF2**

## (b) Electric field strength

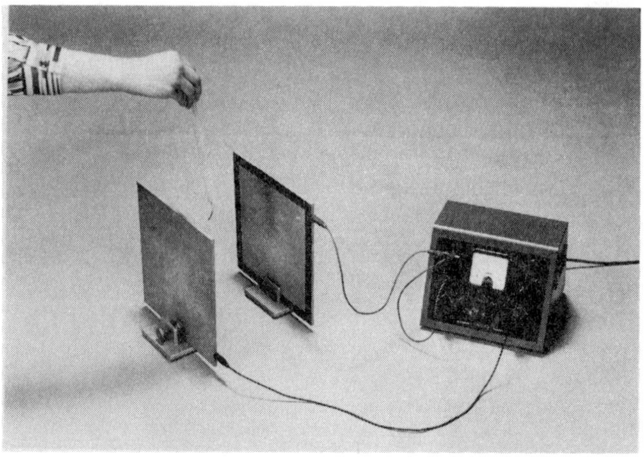

Figure 2.22

### Apparatus

- O capacitor plates (1 pair)
- O 2 slotted bases
- O 2 square polythene tiles
- O polythene strip
- O e.h.t. power supply 0-5 kV
- O aluminium foil

1 Cut a narrow strip (1 mm x 20 mm) of aluminium foil and fasten (by means of adhesive tape) to the polythene strip so that about 10 mm is projecting.

2 Set up the plates as shown in figure 2.22. Arrange the plates about 0.2 m apart and make sure that they are parallel. Adjust the e.h.t. supply to about 2 kV.

3 Charge the foil by contact with the positive plate and place the strip betwen the plates. The foil bends. Why?

4 Use the foil the explore the electric field. Move it sideways, parallel to the plane of the plates. Move it slowly from the negative plate to the positive plate. What do you notice about the deflection of the foil? What does this tell us about the electric field?

5 Place the strip between the plates and increase the p.d. to about 4 kV. What happens to the deflection of the foil? What can you say about the strength of the electric field? Increase the plate separation to about 0.4 m. What can you say now about the strength of the electric field?

## EXPERIMENT FF3

# The potential around a sphere

### Apparatus

- ❍ plastic football (coated with aquadag)
- ❍ nylon thread to suspend the football - or insulating column
- ❍ e.h.t. power supply 0-5 kV
- ❍ VELA with high resistance shunt
- ❍ leaf electroscope with hock
- ❍ flame probe
- ❍ leads

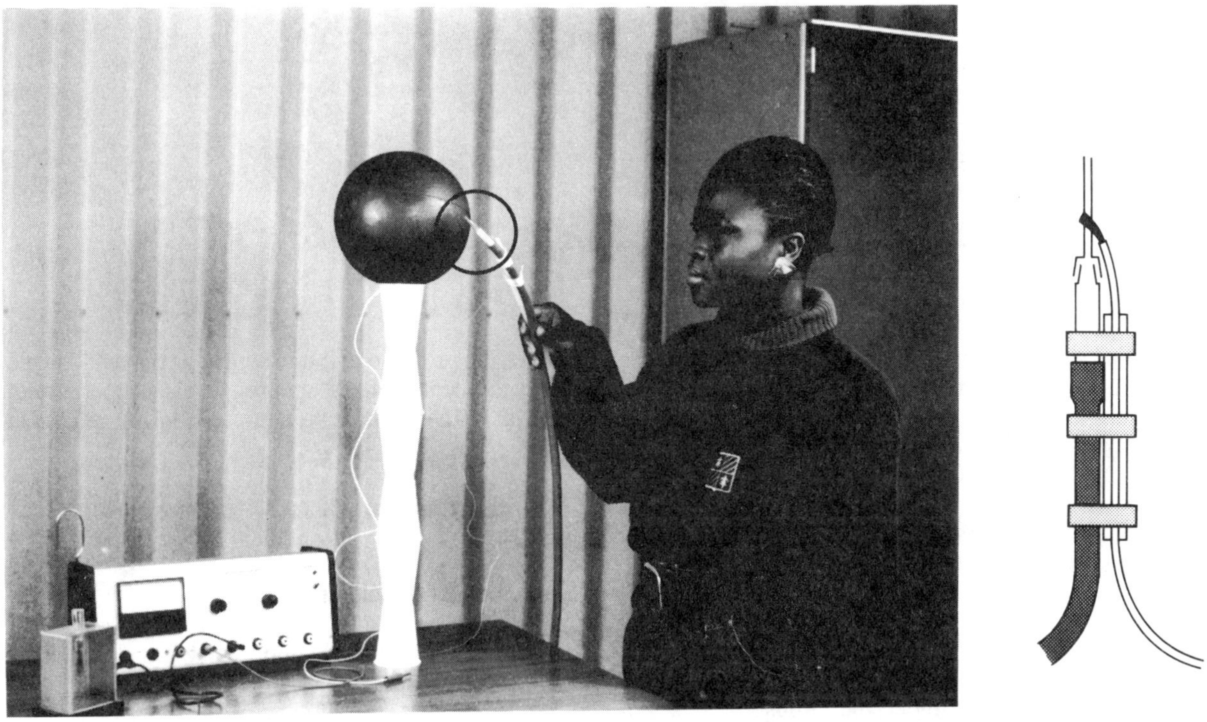

**Figure 2.23**

1 Set up the apparatus as shown in figure 2.23. The sphere should be supported on an insulating column or alternatively hung on nylon thread using a triple suspension to hold it in one place. Keep the sphere away from the walls and bench. The wire which connects the probe and *either* the electroscope *or* the VELA must not touch the bench or any earthed conductor. Make sure that the flame probe is vertical, and that the flame is in the horizontal plane containing the centre of the sphere. Turn the flame out.

**WARNING** Do not move the flame too close to the ball - the heat will melt the plastic.

*Either*

2 Using the leaf electroscope.

The electroscope, which has been calibrated to read potential, should be connected to the probe. It measures the potential at the tip of the probe.

*Or*

2 Using the VELA

The VELA should be connected to the probe. It measures the potential at the tip of the probe. To set up the VELA as a digital voltmeter, press the following keys.

RESET
0 0
ENTER

continued

## EXPERIMENT FF3

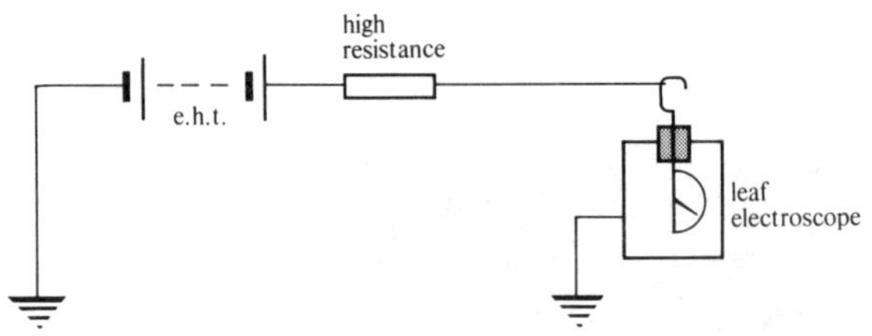


**Figure 2.24**

The reading shown is 100 times less than the true value in kV so you will need to mutliply all readings by 100.

If the VELA is fitted with the Physics EPROM it is possible to set the decimal point to the correct position.

3 With the e.h.t. supply switched off, measure the radius $a$ of the sphere. Switch on the power supply with the probe in contact with the sphere. Record the surface potential.

4 Light the flame and move the probe to a distance away from the centre equal to twice the radius $2a$. Is the potential what you expect?

5 Measure the potential $V$ at different distances $r$ between $a$ and $3a$. Plot a graph that will enable you to deduce a relationship between potential and distance.

6 Move the probe so that it is kept at a constant distance $r$ from the centre of the sphere. What happens to the potential? What shape is the equipotential line around the sphere?

7 Why is it important to keep the sphere away from walls and bench?

### Calibration of the electroscope

1 Connect the e.h.t. power supply to the leaf electroscope, as shown in figure 2.24.

2 Obtain a series of values for the potential difference (as shown by the meter of the e.h.t. supply) in the range 0-1500 V, and the deflection of the leaf.

3 Plot a calibration curve and/or calibrate the scale of the electroscope to read potential.

## EXPERIMENT FF4

# Relationship between charge and potential

### Apparatus

- ❍ e.h.t. supply 0-5000 V
- ❍ plastic football (coated with aquadag)
- ❍ coulombmeter
- ❍ copper wire
- ❍ insulating rod
- ❍ nylon filament

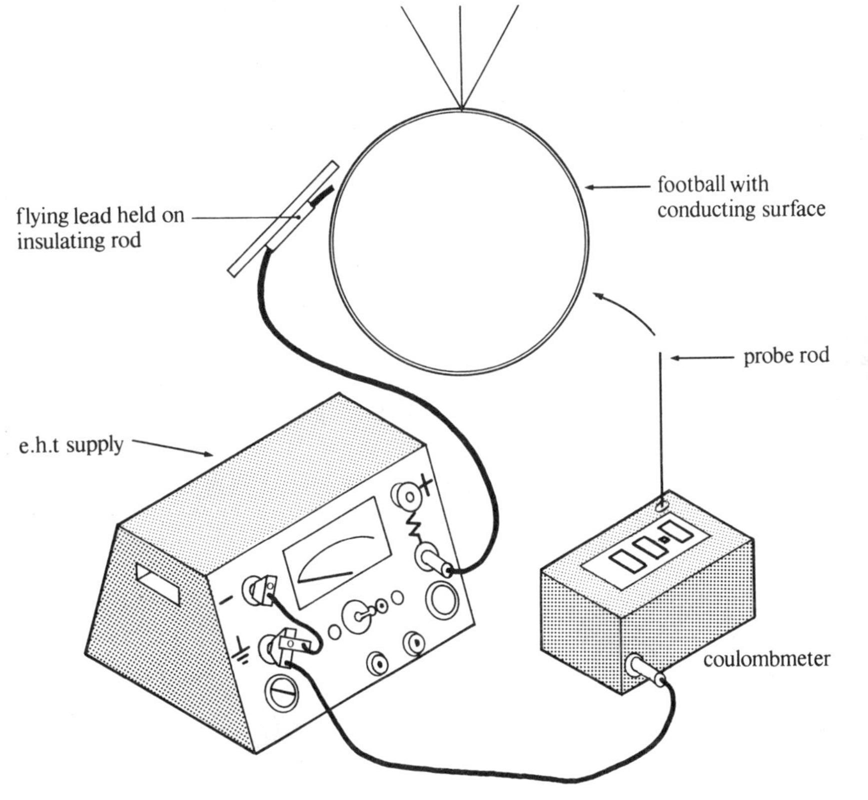


**Figure 2.25**

1 The apparatus is shown in figure 2.25. You must ensure that the flying lead is connected to the positive terminal through a high resistance (1 MΩ). The negative supply is connected to earth and the coulombmeter. The sphere is charged by touching it with the flying lead, which is held by the insulating handle. The flying lead is then removed and the e.h.t. supply switched off. The charge on the ball is measured by transferring it to the coulombmeter by touching the ball with the thick copper wire probe.

2 Measure the radius $a$ of the ball.

3 Set up the apparatus as shown in figure 2.25. Hold the flying lead by the insulating handle and switch on the e.h.t. supply.

4 Measure the charge $Q$ on the ball.

5 Repeat steps 3 and 4 for several values of $V$ and obtain a value for $1/4\pi\varepsilon_0$ using the relationship

$$V = \frac{1}{4\pi\varepsilon_0} \cdot \frac{Q}{a}$$

6 The experiment depends upon nearly all the charge on the ball passing to the coulombmeter. Should the capacitance of the coulombmeter be large or small compared with that of the ball?

7 Estimate what percentage of the original charge stays on the sphere if the coulombmeter has a capacitance of $10^{-2}$ μF. (Use $C = Q/V$ to deduce the capacitance $C$ of the ball.)

*Note.* The concept of capacitance is introduced in the unit *Electronic systems.*

# Questions on objectives

1 On a certain planet a gravitational force of 24 N acts on a body of mass 6.0 kg. What is the planet's gravitational field strength? *(objectives 2 and 12)*

2 Show that the gravitational field strength $g$ at the earth's surface and the gravitational constant $G$ are related by the expression

$$g = \frac{4\pi\rho}{3} GR$$

where $\rho$ is the density of the earth and $R$ is the radius of the earth. *(objective 4)*

3 Sketch graphs to show how
(a) the gravitational field strength,
(b) the gravitational potential due to the earth's mass vary with distance from the surface of the earth (for points outside the earth's surface). State the assumptions that you have made. *(objective 7)*

4 Sketch graphs to show the patterns of electric field lines that result from the following arrangements of electric charges.
(a) a point negative charge,
(b) a positively charged sphere,
(c) two unlike point charges separated by a small distance,
(d) a positively charged horizontal plate, held above and parallel to an earthed plate. *(objective 3)*

5 Define electric field strength and the electric potential at a point in an electric field, and explain the relation between them. *(objectives 2 and 5)*

6 Sketch a graph to show how the electric potential varies with distance on the line between two equal positive point charges. *(objectives 7)*

7 The gravitational potential difference between the surface of the earth and a distance of 50 x $10^6$ m from the centre is 54.7 x $10^6$ J $kg^{-1}$.
(a) How much energy would be needed to move a 2 tonne (1 tonne = $10^3$ kg) spacecraft from the surface to this distance?
(b) If all this energy was supplied as kinetic energy to the spacecraft, what initial velocity would it have? *(objective 12)*

8 Two parallel metal plates are fixed 50 mm apart and are maintained at a p.d. of 2000 V.
(a) Calculate the electric field strength between the plates, assuming that it is uniform.
(b) Calculate the force which would be exerted on a small metallised polystyrene ball, carrying a charge of 4.0 x $10^{-10}$ C, which is suspended between the plates. *(objective 12)*

9 Explain with the aid of diagrams the principles of the van de Graaff generator. *(objectives 10 and 11)*

TOPIC 3

# Magnetic forces and fields

**Study time: 1.5 weeks**

## Summary

When an electric charge moves, there is a magnetic field associated with it. Such a field can be measured by the force it exerts on another current or the couple it exerts on a magnetic dipole (magnet or coil). Our field of study is applied to moving charges and to the design of electric motors and current-measuring instruments.

The photograph shows the *Aurora Borealis* seen from an island in northern Norway. The glow is produced when the upper atmosphere is excited by charged particles (the 'solar wind') funnelled in by the earth's magnetic field.

## Objectives

When you have completed the work in this topic you should be able to:

1 Use the following scientific terms correctly: magnetic field, magnetic pole, neutral point, magnetic field line, permeability of free space, north (seeking) and south (seeking) poles, earth's magnetic field.

2 Define and use the following terms: magnetic flux density, ampere.

3 Recall the standard symbols and units for the following: permeability of free space, magnetic flux density.

4 Describe in words, or by a sketch, the general shape and relative intensities of the magnetic fields around the following current-carrying arrangements:
(a) a long straight wire,
(b) a flat circular coil,
(c) a long solenoid.

5 Investigate experimentally the factors determining the size of the force on a current-carrying conductor in a magnetic field.

6 Determine the direction of the forces due to magnetic fields, given the directions of magnetic field and current.

7 Describe how to measure magnetic flux density, using a Hall probe.

8 Recall, and use to solve problems, the expressions for:
(a) the force on a current-carrying conductor in a magnetic field,
(b) the force on a charged particle in a magnetic field,
(c) the flux density near a long straight wire, at the centre of a circular coil, inside and at the end of a long solenoid.

9 Describe the origin of the Hall p.d., and derive an expression to show how it can be used to measure magnetic flux density, as in the Hall probe.

10 Describe the construction, and explain the principles of operation, of the moving-coil galvanometer and the d.c. electric motor.

11 Solve problems involving the operation of the moving-coil galvanometer.

## Experiments

| | |
|---|---|
| FF5 Force on a current in a magnetic field | (1.5 hours) |
| FF6 Investigation of magnetic flux density using a hall probe | (1.5 hours) |

## References

| | |
|---|---|
| Akrill | Chapter 16 |
| Bolton | Chapter 14 |
| Duncan | Chapter 15 |
| Muncaster | Chapter 41 |
| Nelkon | Chapters 10 and 11 |
| Wenham | Chapters 27and 32 |
| Whelan | Chapters 52, 53 and 54 |

## 3.1 Magnetic fields

In the previous topics the concept of a field has been developed to describe a region where forces act 'at a distance'. The field was mapped by field lines which showed the forces acting on a charge in an electric field and on a mass in a gravitational field.

There is a third type of 'action at a distance' with which you are already familiar - that between magnets. We will discover in this topic that this magnetic interaction occurs between any electric charges which are moving relative to each other. The force exerted by magnets is one important example of this effect.

### Fields due to magnets

The magnetic field around a magnet is a region in which forces act on other magnets and on magnetic materials, like iron, which become magnetised in this field. The main properties of magnets have been known since at least 600 BC. The magnetic properties of a magnet seem to originate in two regions we call **poles.**

#### Q 3.1 Study question

Make three brief statements about the properties of magnets which each include the word 'pole'. ■

The strength and direction of a magnetic field can be represented by **magnetic field lines.**

The direction of the magnetic field at a particular point can be defined in several ways, but we will begin by using the one which can be easily applied in practice.

The direction of the magnetic field is the direction in which the **north** pole of a small compass needle will point when it is freely suspended through its centre of gravity. When a magnetised compass needle is placed in a magnetic field (figure 3.1) it will be acted on by two equal and opposite forces *F* which form a couple. This couple will give it an angular acceleration and, if there is damping, it will eventually set aligned with the field.

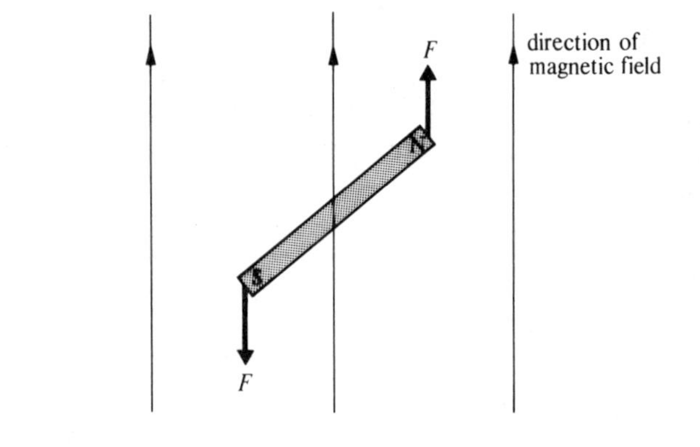

**Figure 3.1**

#### Q 3.2 Self-assessment question

Why does a compass needle mounted on a vertical pivot respond only to the horizontal component of the magnetic field? ■

Compared with the magnetic fields we shall be investigating in this topic, the earth's field is very weak. Near to magnets and electric currents we can usually ignore the effects of the earth's magnetic field although, at greater distances from the magnets or currents, we may have to take the earth's field into account.

#### Q 3.3 Self-assessment question

(a) Explain briefly what is meant by (i) a magnetic field line, and (ii) a neutral point.
(b) Write a brief statement answering each question below and giving a reason for your answer.
(i) Do magnetic field lines have a beginning and end? (ii) Can magnetic field lines intersect? (iii) How is a uniform magnetic field represented? (iv) Can a magnetised compass needle which produces its own magnetic field be used to investigate the field around a small bar magnet? ■

### Fields due to electric currents

Although the properties of magnets were known to early Greek scientists and by 1600, William Gilbert the Elizabethan court physician, had proposed the idea that the earth was itself a magnet, it was not until 1821 that Oersted, a Danish schoolteacher, discovered the magnetic field around an electric current.

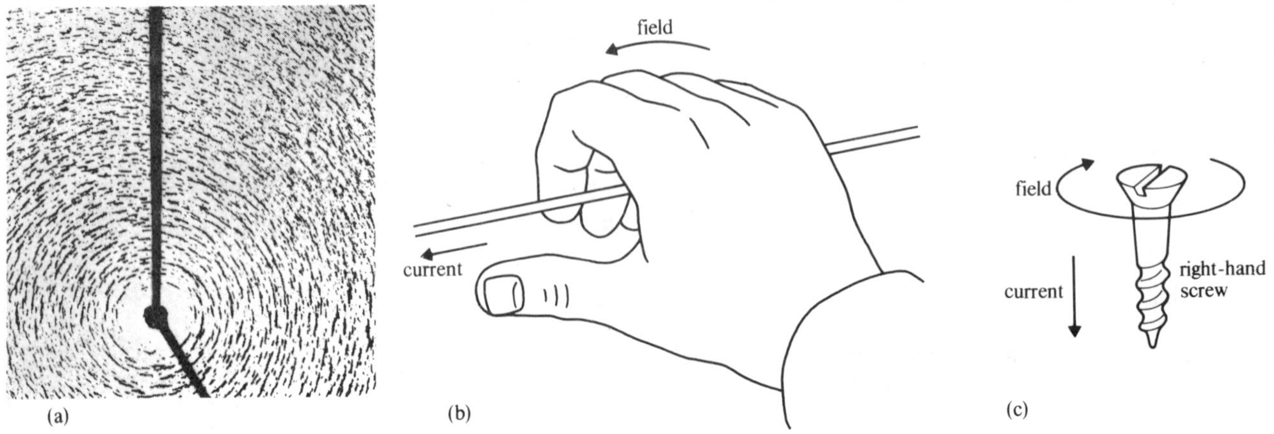

**Figure 3.2**

In 1878, Rowland demonstrated that a magnetic field could be produced around an insulated charged disc rotated at high speed. This showed, more fundamentally, that electric charges in motion are the cause of magnetic fields. We must presume that a field produced by magnetised materials is due to the movement of electrons within the atoms of the material.

Figure 3.2(a) shows the iron-filing field pattern around a straight wire. Figures 3.2(b) and (c) illustrate two rules for predicting the direction of the field.

### Q 3.4 Self-assessment question

State in your own words two rules which give the direction of the field around a current. Use one of the rules to sketch a field line around

(a) a current flowing out of your paper (represented by the sign, (•)) indicating the point of an approaching arrow), and
(b) a current flowing into your paper (represented by the tail of a receding arrow, ⊗ )■

### Q 3.5 Self-assessment question

Figure 3.3 is a photograph taken from above a horizontal board showing an iron filing field pattern due to a current in a circular wire loop.

(a) Sketch the field produced in a *vertical* plane through line $Y_1$ $Y_2$ if the current goes from left to right along the top of the loop. Indicate the direction of current and field and say whether you are viewing the pattern from point $X_1$ or point $X_2$.
(b) Can you apply the right-hand screw rule to the loop even when field and current representations are interchanged? Illustrate your answer. ■

Figure 3.4 shows the separate horizontal magnetic fields produced by the current flowing in the two opposite sides of a vertical loop.

The circular fields indicate that, at P, the compass needle will be acted on by two equal strength component fields and, as a result, will set along the axis of the loop. Point Q is nearer to one side of the loop and the compass will be more strongly influenced by the current on one side. To find the direction of this resultant field at Q (or any other point), we use the parallelogram law for vectors to add up contributions. The field photographed in figure 3.3 is the resultant field. (The field of the earth is weak enough to be ignored.)

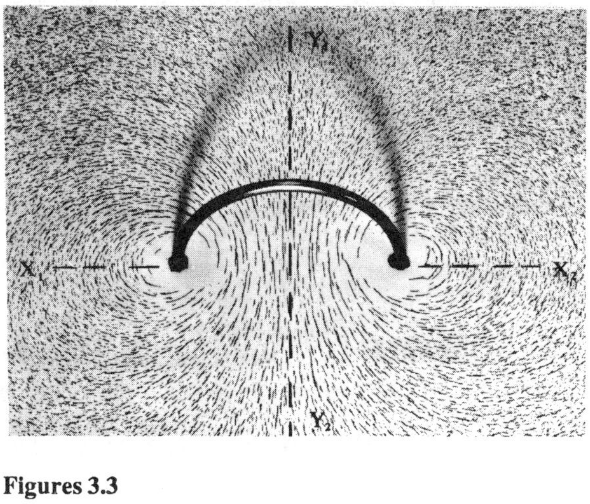


Figures 3.3

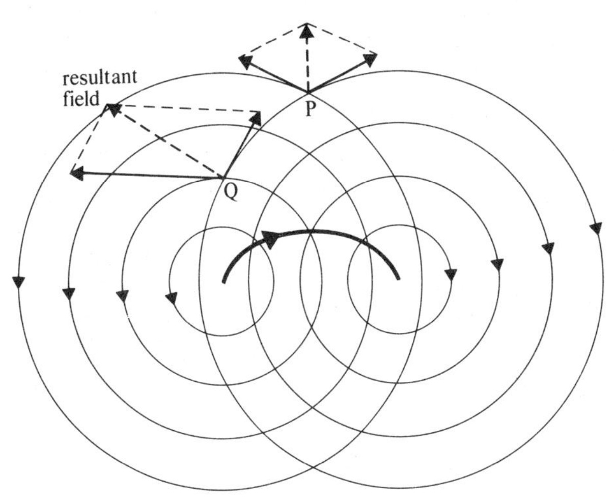


Figure 3.4

### Q 3.6 Study question

What is a solenoid?

Sketch the magnetic field around a solenoid, marking the direction of current and magnetic field lines. Say how your two- dimensional diagram relates to the complete three-dimensional field.

Why is it possible, and often preferable, to use a solenoid in place of a bar magnet? ■

### Q 3.7 Self-assessment question

A long vertical wire carrying a current passes through a horizontal board. Sketch the resulting magnetic field on the board due to the current and the earth's magnetism, assuming the field due to the current decreases with distance from the wire and the earth's magnetic field is uniform. Mark the position of any neutral points formed in this field. ■

### Q 3.8 Discussion question

We could not learn, reason and predict without pattern. Within pattern, symmetry is important. An inverse square law follows from the symmetrical spread of field round a charge; remove one half of a long solenoid and the field at the new end must be half of what it was.

There is symmetry of charge (electron and positron) and of matter and anti-matter. Why should not a single magnetic pole (monopole) exist? What would its properties be? ■

**Background reading**
*New Scientist*, 16th October 1986, p.53 and 17th September 1987, p.49.

## 3.2 Forces in magnetic fields

You have observed that both magnets and electric currents produce magnetic fields. These fields were detected by the effect of the forces on a small magnet. Since a current-carrying conductor establishes a magnetic field, can we expect that a current carrier will also experience a force in a magnetic field?

You can now investigate whether magnetic fields can be detected and measured by the force exerted on a current-carrying conductor.

## E Experiment FF5
**Force on a current in a magnetic field**

In this experiment, you will investigate the direction of the force exerted on a current-carrying conductor in a magnetic field, identify the factors which determine the magnitude of this force and so find a way of measuring a magnetic field.

## Direction of the force

The results of the first part of experiment FF5 show that the direction of the force on the current-carrying conductor is *perpendicular* to the directions of the field and the current (figure 3.5).

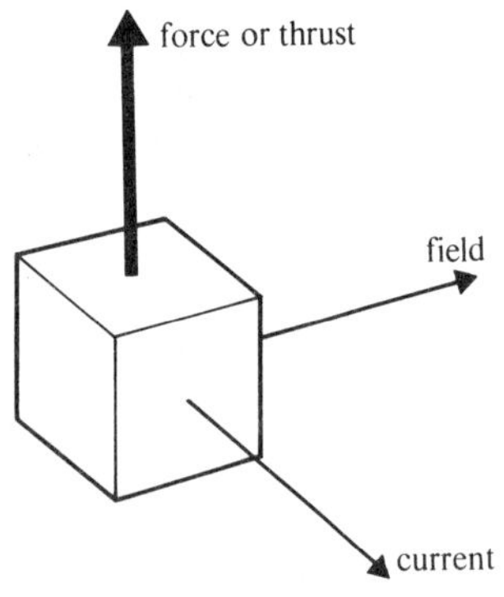


**Figure 3.5**

The particular direction of the force, when the directions of the current and field are known, can be found using Fleming's left-hand rule (figure 3.6).

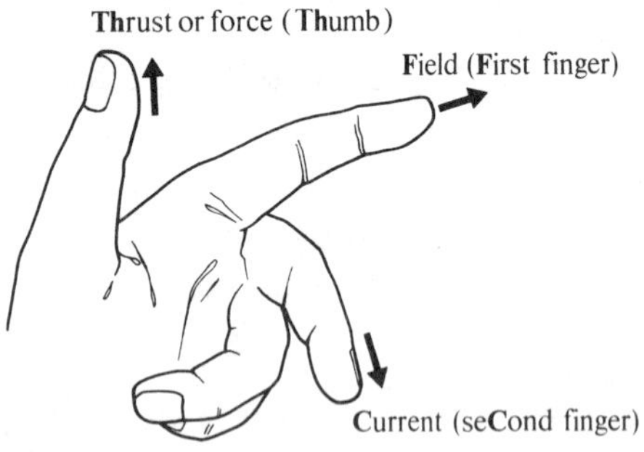


**Figure 3.6**

*Note.* The word thrust is used for the force exerted so that it can be linked to the thumb by the letters th.

### Q 3.9 Self-assessment question

Which of the arrangements shown in figure 3.7 will produce a force to tilt the current balance? ■

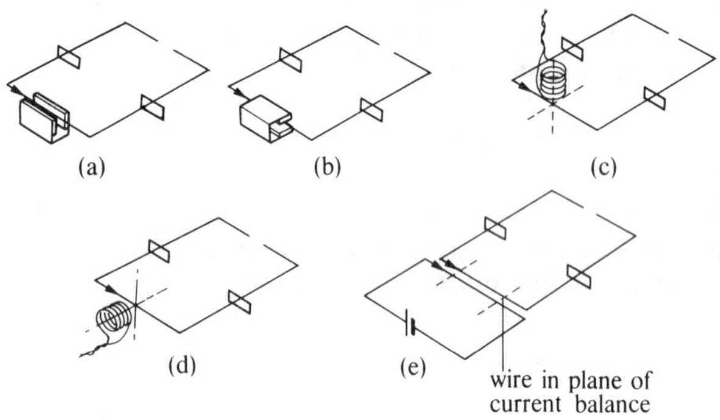


**Figure 3.7**

### Q 3.10 Development question*

In figure 3.8 the magnetic field due to a current (in a direction 'out of the paper') is shown. The field lines due to the magnet are not shown. Magnetic field lines indicate the direction of the force which acts on the N-pole of the magnet.

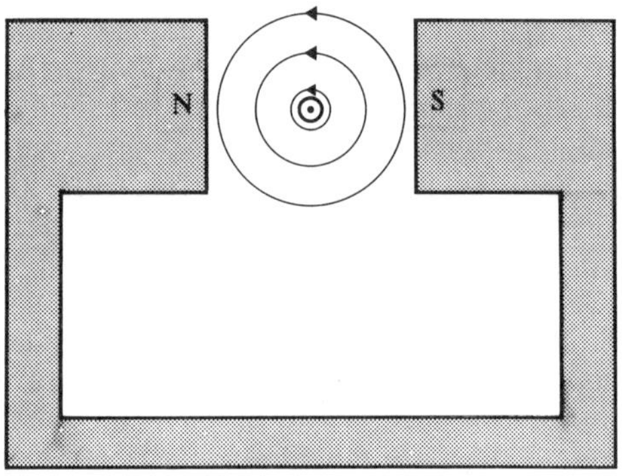


**Figure 3.8**

(a) Sketch figure 3.8 and mark in the direction and size of the forces acting on the poles of the magnet due to the current if the total force on the magnet due to the current is $F$.
(b) Since forces always act in 'third law pairs' there must be a force on the current-carrying wire due to the magnetic field of the magnet. Sketch the field lines due to the magnet and show the force on the wire. Does your diagram agree with Fleming's left-hand rule?
(c) Sketch the resultant magnetic field due to current *and* magnet and mark the forces acting on wire and magnet. If the current-carrying wire was free to move, would it accelerate towards the region of stronger or weaker magnetic field? ■

### Q 3.11 Self-assessment question

What is the direction of the force on segments A and B of the coil (figure 3.9) when the current $I$ creates a magnetic field through the coil? What is the likely effect of switching on a huge current of 1000 A in the coil if ways are found of keepng the coil cool? ■

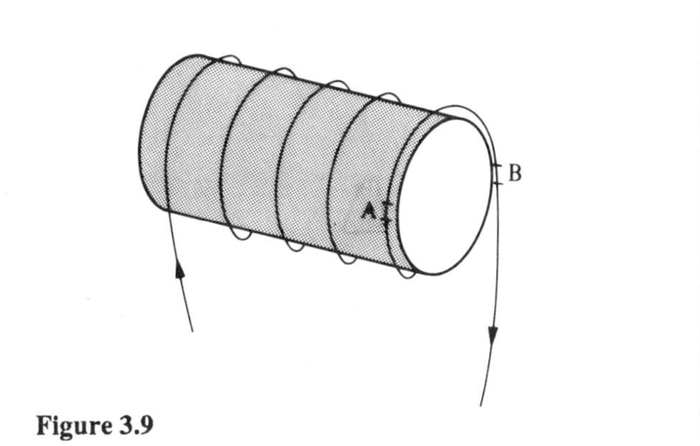


Figure 3.9

The parallel with electric and gravitational fields is clear since the magnitude of an electric field is measured by the ratio of the force to the charge acted on, and a gravitational field by the ratio of the force to the mass acted on (figure 3.10).

*Note*: $B$ is *not* called called field strength; the quantity $B$, magnetic flux density (also called **magnetic induction**), represents the magnitude, direction and sense of the magnetic field. Sometimes, the term ***B*-field** is used - a shorthand way of describing **a magnetic field of flux density *B*.**

### Q 3.12 Self-assessment question

Use the equation $B = F/Il$ to derive the dimensions of flux density and write a definition of the tesla. ■

## 3.3 Magnetic flux density

Experiment FF5 demonstrated that the force $F$ on a conductor of length $l$ carrying a current $I$ perpendicular to a magnetic field is determined by the following relationships.

$$F \propto I \text{ (length constant)}$$

$$F \propto l \text{ (current constant)}$$

So $$F \propto Il$$

The force was also observed to depend on the magnitude of the magnetic field, a fact which can be used to *define* the magnitude of the field by introducing the term **magnetic flux density *B*.**

Magnetic flux density is defined by the equation

$$F = BIl$$

Magnetic flux density is numerically equal to the force acting per unit length of a conductor which carries unit current at right angles to the direction of the field.

$$B = \frac{F}{Il}$$

The unit of $B$ is the newton per ampere-metre (N $A^{-1}$ $m^{-1}$) which is given the name **tesla (T)**.

Magnetic flux density is defined by a ratio - the ratio of a force to the thing being acted upon (a length of current perpendicular to the field).

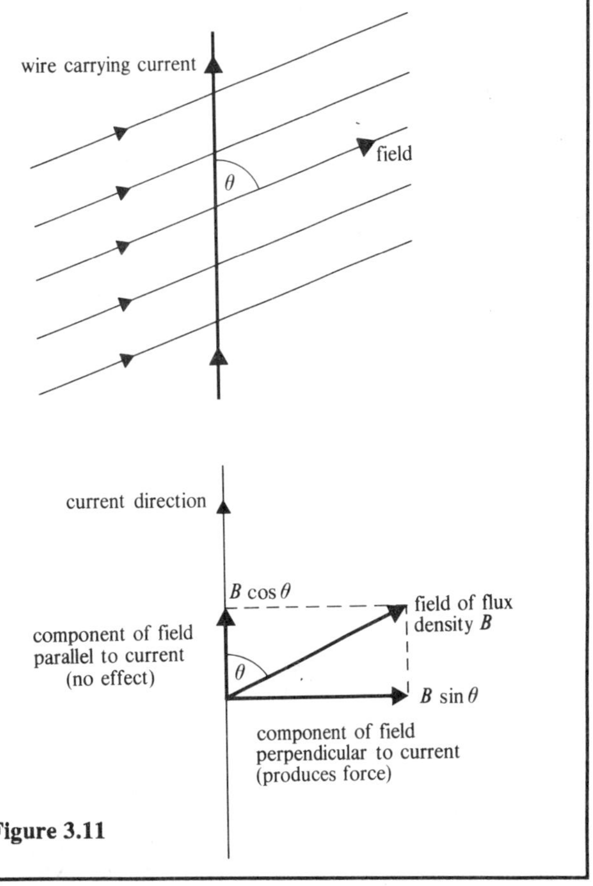


Figure 3.11

Experiment FF5 showed that there is no force when the current direction is parallel to the direction of the magnetic

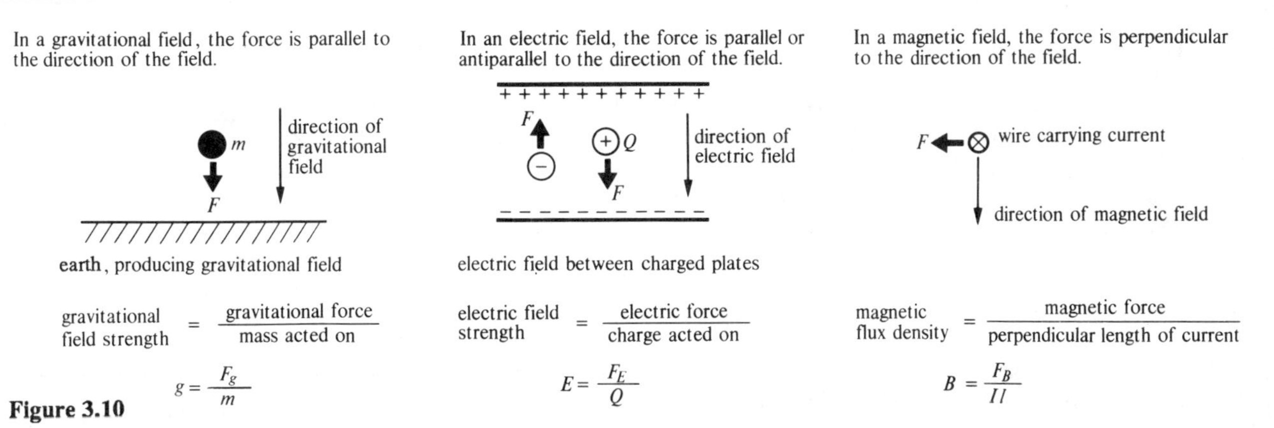


Figure 3.10

field. When the directions of the current and the field are at some other angle (figure 3.11) the resolved component of the field perpendicular to the current direction $B \sin \theta$ produces the force (while the component parallel to the current $B \cos \theta$ produces no effect).

Then $$F = BIl \sin \theta$$

How is the representation of a field by magnetic field lines related to flux density?
(i) The tangent to a field line at any point gives the direction of the flux density $B$ at that point. (ii) The magnitude of flux density $B$ is represented by the 'density' of the field lines. The number of field lines passing through unit cross-sectional area perpendicular to the field is proportional to the magnitude of $B$. A uniform field is indicated by evenly spaced lines.

The term **magnetic flux** will be defined in the unit *Electromagnetism* but it may help you to know now that magnetic flux is measured in the unit **weber** (Wb) so flux density can be measured in weber per metre squared.

$$1 \text{ Wb m}^{-2} = 1 \text{ T}$$

## Q 3.13 Self-assessment question

One arm of a current balance, of length 25 cm, is placed inside a flat solenoid (see figure 3.22f in experiment FF5). The solenoid and balance are connected in series and when there is a current of 1.4 A in each, the balance must be counterpoised by placing a rider of mass 0.12 g on the arm of the balance an equal distance on the other side.
(a) Calculate the flux density inside the solenoid when the current is (i) 1.4 A, (ii) 1.0 A.
(b) What current would produce a flux density inside the solenoid equal to the flux density $B_H$ due to the horizontal component of the Earth's magnetic field if $B_H = 1.8 \times 10^{-5}$ T? ■

# 3.4 Force on a moving charge

An electric current is produced by the drift of charged particles (for example, electrons in metals, charged ions in electrolytes and gases). The magnetic force you have observed acting on a current-carrying conductor must be the resultant of all the forces acting on these charged particles as they move through a magnetic field. Let us use this assumption to derive an expression for the force on a single charged particle moving in a magnetic field.

## Q 3.14 Development question*

Figure 3.12 shows a magnetic field of flux density $B$ passing at right angles through length $l$ of a conductor.
Suppose the conductor contains $N$ charge carriers in length $l$ between X and Y. Each carrier has charge $Q$ and is moving with velocity $v$.
(a) If each charge carrier takes time $t$ to travel from X to Y express $l$ in terms of $v$ and $t$.
(b) What is the total charge passing Y in time $t$?
(c) What current flows through the field?
(d) What is the force on all $N$ charges in the field (assuming $F = BIl$)?
(e) What force acts on each charge $Q$?
(f) If the charge-carriers in figure 3.12 are electrons, what is the direction of current?
(g) What is the direction of the force on each electron? ■

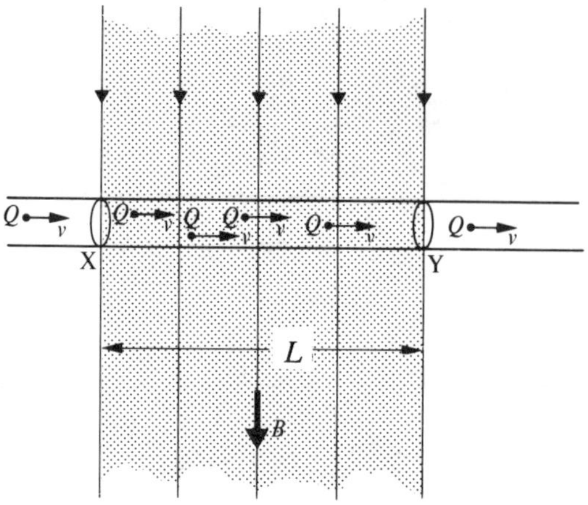


Figure 3.12

*Note*. In this topic, flux density $B$ has been defined from the equation $B = F/Il$, and we have used this equation to show that $F = BQv$ for a charge $Q$ moving with velocity $v$ at right angles to $B$. Alternatively, $B$ can be defined by the equation $F = BQv \sin \theta$ and this equation can then be used to derive $F = BIl \sin \theta$, where $\theta$ is the angle between the direction of the current (or charge carriers) and the magnetic field.

One of the important effects of the force which acts on a charge moving in a magnetic field is called the **Hall effect.** It will be convenient to discuss the Hall effect at this point because it provides a useful means of measuring the flux density.

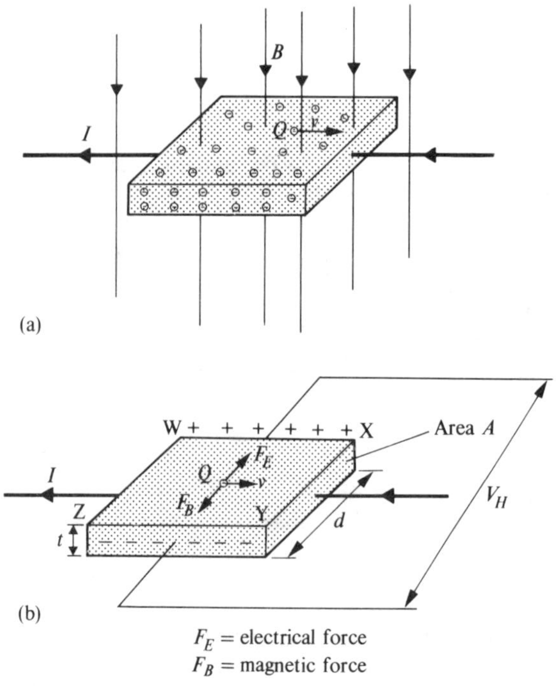


Figure 3.13

Figure 3.13 represents a slab of conductor carrying a current

$I$ produced when carriers of *negative* charge $Q$ drift with velocity $v$ *in a direction opposite to I*. In the following question you will use an equation, developed in topic 5 of the unit *Behaviour of matter*, relating current $I$ and drift velocity $v$. If there are $n$ current carriers per unit volume the current $I = nQAv$ (since all charge carriers moving across any cross-section in one second will be those contained in volume $vA$, i.e. $nvA$ carriers crossing per unit time each carrying charge $Q$).

## Q 3.15 Development question*

(a) If a magnetic field of flux density $B$ is applied as shown in figure 3.13a, what is the direction of the magnetic force on the moving particles carrying negative charge $Q$ ?
Charge-carriers will be accelerated in the direction of this force producing a higher density of negative charge-carriers near the front face and a shortage towards the back (figure 3.13a). An electric field, $E$, will build up across the conducting slab like the field between charged plates until the extra negative charges stop building up at the front, because the electric force on the moving charges balances the magnetic force (figure 3.13b).
(b) State the force acting on a charge $Q$ due to electric and magnetic fields and express the balance shown in figure 3.13b.
(c) There will be a potential difference $V_H$ between front and back faces. Write an equation expressing the fact that the field strength $E$ is equal to the potential gradient.
(d) Write an equation for $V_H$ in terms of $B$ using your answers to (b) and (c).
(e) Hence show that $V_H = BI/nQt$ where $t$ is the slab thickness. ■

The Hall effect produces a potential difference $V_H$ *across* a current-carrying conductor in a magnetic field. Of course, there is also a p.d. *along* the conductor which produces the current $I$. In figure 3.13b the direction of current $I$ indicates that Y is at a higher potential than W or Z. However, Y and X may be at the same potential when the $B$-field is absent but when this field is applied the potential of X will be raised above Y in this particular case. To measure $V_H$ a sensitive meter is placed between two points such as the mid-points of sides WX and YZ. These two points are at the same potential when the current is on but no magnetic field is applied. The $B$-field then creates a Hall p.d. across the meter.

The equation $V_H = BI/nQt$ indicates that the Hall p.d. is proportional to $B$ and so the effect can be used to devise a method for comparing flux densities.

## Q 3.16 Study question

(a) Explain briefly using your reference texts and the expression for $V_H$ why the Hall effect is more marked in semiconductors than in metals.
(b) Explain how the Hall effect can be used to determine the sign of the majority charge-carriers in a conductor or semiconductor.
(c) What is a Hall effect proximity switch? ■

## Q 3.17 Self-assessment question

Figure 3.14 shows a copper slice carrying a current of 9.0 A. When a magnetic field of 0.75 T is applied as shown, a Hall p.d. of 3.8 μV is produced between points X and Y.

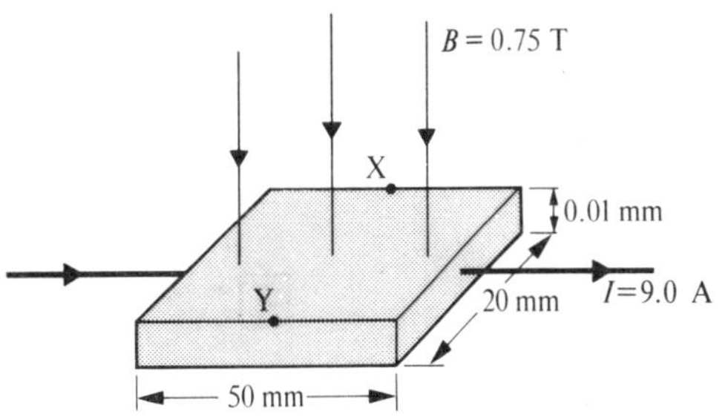


**Figure 3.14**

(a) How many free electrons per unit volume are there in copper according to these results?
(Magnitude of charge on electron = $1.6 \times 10^{-19}$ C)
(b) Calculate the number of atoms per unit volume of solid copper if this particular slice of copper has a mass of 0.89 g and 1 mole of copper ($6.0 \times 10^{23}$ atoms) has a mass of 64 g.
(c) How many free electrons per atom are there in this copper according to the results given?
(d) Calculate the drift velocity of the free electrons in the copper slice when the current is 9.0 A. ■

## 3.5 Measurements of magnetic flux density

Three methods can be listed for measuring the magnitude of $B$ or for comparing different values of $B$.

1 A current balance.
2 A Hall probe.
3 A search coil with an oscilloscope for detecting the variable fields produced by an alternating current.

In method 1 a balance like that used in experiment FF5 could be used and the force, $BIl$, measured directly using counterweights. In practice, this method is not sensitive enough.

Method 3 is discussed in detail in the unit *Electromagnetism* because it depends on the induction effects produced when the flux-linkage changes. An oscilloscope connected to a search coil can detect the variable e.m.f. induced in the search coil when it is placed in a varying field (e.g. the field inside a solenoid conducting an alternating current). The oscilloscope provides a convenient method of measuring the peak value of a variable magnetic flux density and is one of the experiments suggested in the unit *Electromagnetism*.

The Hall probe uses the Hall effect to compare magnetic flux densities. The probe consists of a thin slice of semi-conductor mounted on an arm so that the slice can be placed in the magnetic field with the direction of $B$ at right angles to the largest faces of the slice.

The probe is used with a cell, circuit box and galvanometer (figure 3.15). There is no need to memorise any details of the circuit box but note that (i) the cell provides a current through the slice since the Hall effect depends on charges drifting through the field, and (ii) there may be a small p.d. between points G and R generated by the current through the slice and so a balancing potentiometer is required to ensure that the galvanometer reading is zero when the probe is not in the magnetic field.

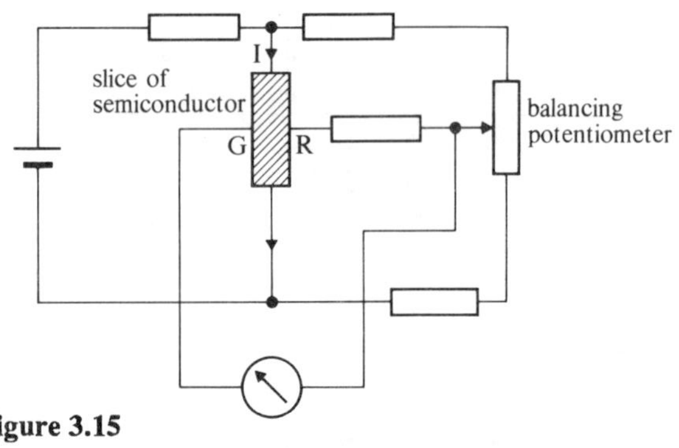

**Figure 3.15**

The probe is used for comparing flux densities so there is no need to know the density of charge carriers in the slice or the calibration constant of the galvanometer.

---

## E Experiment FF 6

**Investigation of magnetic flux density using a Hall probe**

In this experiment, the probe is used to investigate how the magnetic flux density varies inside and around different current-carrying conductors (e.g. solenoids, circular coils and straight conductors).

---

## Calculating flux density

Your experimental observations have indicated the important factors which determine the flux density produced by different current-carrying conductors. Not surprisingly in every case, flux density $B$ is found to be proportional to the current. Other important conclusions are summarised below.

A **solenoid** has a flux density in the *inner region* which is constant over the whole cross-section (i.e. on the axis or away from the axis) and if the solenoid is long (i.e. its length is great compared with its cross-section) $B$ stays constant over a large part of its length.

The field at the *centre* of an *end* face is half as big as the field well inside the solenoid (and this field is not uniform across the end face).

The field does not depend on the shape of the cross-section or the cross-sectional area provided the solenoid is 'long'. Your experiments also showed that the flux density $B$ inside a long solenoid which has $N$ turns and length $l$ is proportional to the current, and to the number of turns per unit length.

$$B \propto \frac{NI}{l}$$

The flux density also depends on the material filling the solenoid. If this space is empty (it makes virtually no difference whether it is a vacuum or air) we can write

$$B = \frac{\mu_0 NI}{l}$$

$$B = \mu_0 nI$$

where $\mu_o$ is a constant called the **permeability of free space** (or vacuum) and $n$ is the number of turns per unit length.

### Q 3.18 Self-assessment question

Show that a suitable unit for permeability is T m $A^{-1}$ and express this unit in terms of the newton and the ampere. ■

As we shall see later in this topic the permeability of free space $\mu_o$ is fixed by the definition of the ampere as $4\pi \times 10^{-7}$ N $A^{-2}$.

You know that the flux density inside a solenoid can be greatly increased by filling the solenoid with a core of ferromagnetic material (e.g. soft iron) which itself becomes magnetised.

Then $B = \mu n I$ where $\mu$ is the permeability of the material of the core and is not a constant but depends on the state of magnetisation of the core. (This topic is studied further in the unit *Electromagnetism*.)

**A flat circular coil** has a flux density which varies across its face but at the centre

$$B = \frac{\mu_0 NI}{2a}$$

where $a$ is the radius and $N$ the number of turns.

**A long straight conductor** produces a field with a flux density which varies inversely as $r$, the perpendicular distance from the wire.

$$B \propto \frac{I}{r} \quad \text{or} \quad B = \frac{kI}{r}$$

Experimental measurement confirms that k = $2.0 \times 10^{-7}$ N $A^{-2}$ which is $\mu_o /2\pi$.

For a long straight wire

$$B = \frac{\mu_0 I}{2a}$$

You need to memorise the above results and be able to use them to calculate the flux density in different situations.

### Q 3.19 Self-assessment question

Calculate the flux density $B$ in the following.

(a) Inside a solenoid of 1200 turns, length 0.75 m, diameter 50 mm carrying a current of 5.0 A.
(b) At the end of the same solenoid, with the same current.
(c) 1.0 cm from a long straight conductor carrying a current of 5.0 A.
(d) At the centre of an open circular coil of radius 5.0 cm, with 100 turns, carrying a current of 5.0 A.
($\mu_o = 4\pi \times 10^{-7}$ N $A^{-2}$) ■

### Q 3.20 Self-assessment question

The arrangement in figure 3.16 shows a conductor in two

(a) What contribution to the field at the point P is made by the straight wires?

(b) Calculate the total field at P which results from this arrangement. ■

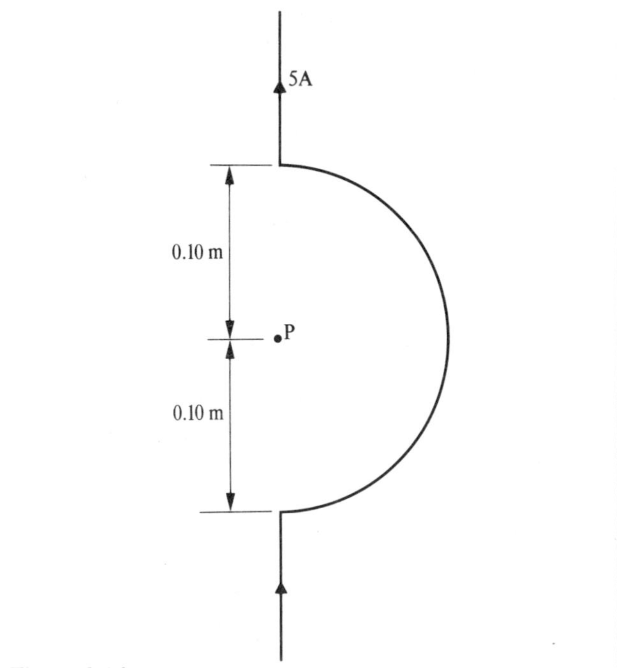

Figure 3.16

## Q 3.21 Self-assessment question

It is necessary to perform an experiment in a small region where the effect of the earth's magnetic field is to be eliminated. The centre of a flat circular coil of radius 0.60 m and 160 turns is chosen as the experimental site.

(a) Sketch how the coil must be orientated and indicate the current direction if the direction of the earth's field is north and pointing 70° down from the horizontal.
(b) What current is required in the coil to cancel the earth's field which has flux density $5.2 \times 10^{-5}$ T at this place? ■

# 3.6 Measurement of current

You have observed that both force on a current-carrying conductor and the magnetic flux density produced by a current are directly proportional to the size of the current. The magnetic effect of a current thus provides a way of defining a unit for the measurement of current.

## The force between currents

Consider now the forces acting in the magnetic field created by two straight infinitely long conductors carrying currents in the same direction.

## Q 3.22 Development question*

Figure 3.17 shows two long parallel straight conductors in vacuum (or air) carrying currents $I_1$ and $I_2$ distance $a$ apart. The currents create a magnetic field throughout space but the diagram shows only the field near the right-hand conductor due to current $I_1$.

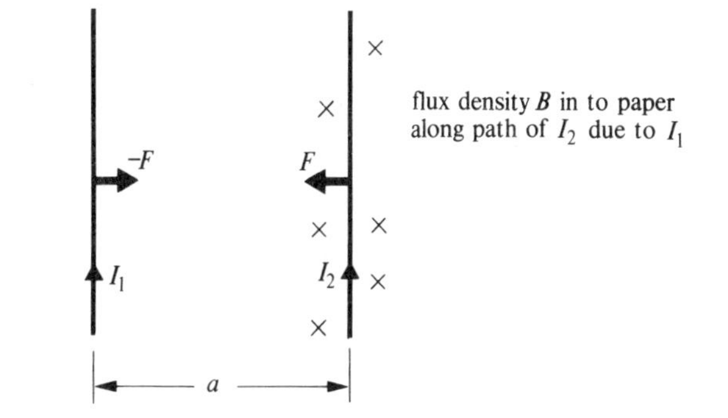

Figure 3.17

(a) Write down an expression for the flux density $B$ at any point along the path of $I_2$ due to the current $I_1$.
(b) Derive an expression for the magnitude of the force $F$ acting on length $l$ of the right-hand conductor carrying current $I_2$ through the field due to the current $I_1$.
(c) Explain the origin of the force $-F$, giving two reasons why it must have the magnitude and direction shown.
(d) Use the expression derived in part (b) to deduce the force which will act on 1 m of either conductor if each conductor carries a current of 1 A and the separation $a$ is 1 m. ■

## Q 3.23 Study question

Look up and then write down the definition of the ampere. Use this definition and your answer to question 3.22(d) to find the value of $\mu_o$.

The constant $\mu_o$ is the constant relating the magnetic flux density in free space to the current which produces this flux, and its numerical value will obviously depend on the unit in which current is measured.

The definition of the ampere determines that $\mu_o$ is $4\pi \times 10^{-7}$ N A$^{-2}$. Experimental measurements of flux density $B$ in current-carrying solenoids will confirm this value for $\mu_o$. The results are not measuring $\mu_o$ but are really saying that the ammeters are correctly calibrated: $\mu_o$ is a defined constant called the **magnetic constant** or **permeability of free space**. By contrast $\varepsilon_o$, the permittivity of free space, is a constant which has no defined value but must be found by experiment. Electromagnetic field theory also predicts that the speed of electromagnetic radiation in free space is given by

$$c = \frac{1}{\sqrt{(\mu_0 \varepsilon_0)}}$$

(see topic 4 of the unit *Waves and vibrations* ) and this relationship can be used to provide an accurate measurement of $\varepsilon_o$.

## Q 3.24 Study question

What is meant by an absolute measurement of current? How does it differ from other methods of measuring currents? Outline briefly an absolute measurement of current. ■

## Q 3.25 Self-assessment question

Show that the equation below is dimensionally correct.

$$c = \frac{1}{\sqrt{(\mu_0 \varepsilon_0)}}$$ ■

## Q 3.26 Self-assessment question

(a) Draw a diagram showing the magnetic field pattern in a horizontal plane due to two long vertical wires P and Q carrying equal currents in the same direction.
(b) Mark on your diagram points X and Y along a line passing through the wires; X is between the wires and twice as far from P as from Q; X and Y are equidistant from Q. Mark on your diagram vectors representing the component $B$-fields at X and Y due to each current.
(c) If $B$ is the flux density at X due to the current in Q, what is the resultant flux density at X and Y?
Does the force acting on each wire tend to make it move towards or away from a region of weaker magnetic field?
(d) Predict, with reasons, the direction of forces acting on parallel conductors carrying currents in *opposite* directions. How does the field vary around one of these conductors at points equidistant from the conductor? Does the force acting on the wire move it towards or away from the weaker field? Illustrate your answer with a sketch of the field. ■

## Moving - coil galvanometers

Absolute methods of measuring current involve direct measurement of the forces acting between current-carrying coils using instruments called current balances. These instruments provide standards against which a wide range of practical forms of current meters can be calibrated. The most common is the moving-coil instrument. This consists of a coil which is either suspended or pivoted in a uniform radial magnetic field. The torque on the coil is directly proportional to the current. This is developed in the following question.

## Couple on a coil in a magnetic field

## Q 3.27 Development question*

Figures 3.18(a) and (b) show a rectangular coil of $N$ turns carrying current $I$ in a uniform magnetic field. $\theta$ is the angle between the direction of the field and the normal to the plane of the coil.

(a) Sketch figure 3.18(a) and mark the direction and magnitude of the forces on sides BC and AD.
(b) Does a force act on side AB and side CD? If so, what are the directions of these forces, and their resultant effect?
(c) Show that a couple of torque $T$ acts on the coil, given by the equation

$$T = BIAN \sin \theta$$

For what position of the coil is the torque maximum? ■

Figure 3.19 shows a rectangular coil, two sides of which are in a radial magnetic field in the space between cylindrical concave pole pieces and a soft iron cylindrical core. If the coil rotates in this field its plane will always be parallel to the magnetic field lines (i.e. $\theta$ will always be 90°) and the torque will be $BIAN$ in all positions.

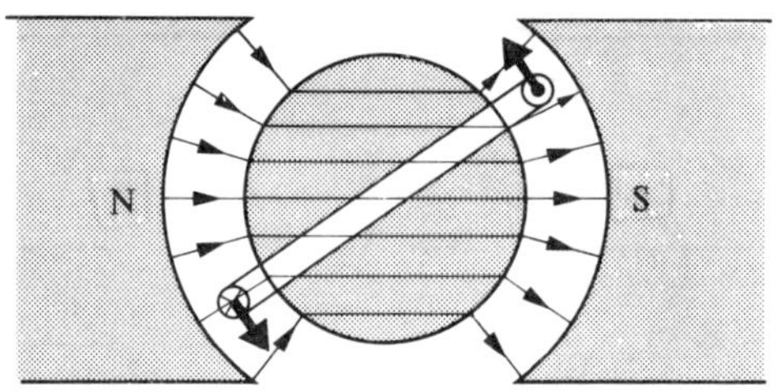


**Figure 3.19**

## Q 3.28 Study question

(a) Describe, with the aid of a labelled diagram, the construction and operation of a moving-coil galvanometer. Explain how a steady deflection can be produced and continuous oscillation of the coil prevented.
(b) Derive an equation for the current $I$ in terms of the deflection $\theta$ which it produces. Explain why the scale is linear. ■

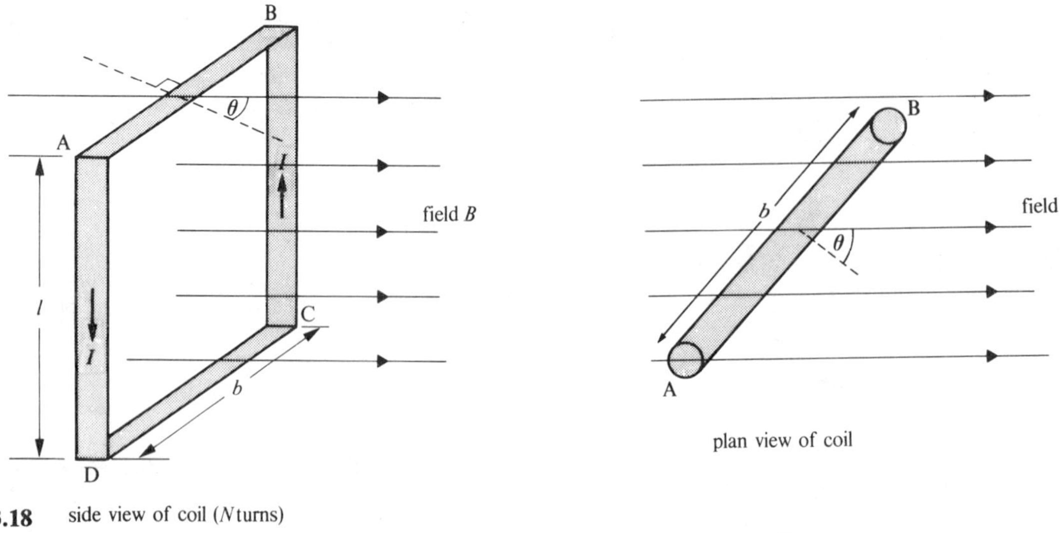


**Figures 3.18**

The current sensitivity of a galvanometer is the ratio of the deflection produced to the current which produces it.

$$S = \frac{\theta}{I}$$

It is measured in appropriate units, e.g. radians per milliampere, division per microampere.

Sometimes, the reciprocal of sensitivity is indicated on a galvanometer, e.g. 2 μA per division, which indicates the smallest current change that can be reliably detected by the galvanometer.

## Q 3.29 Study question

(a) Account for the factors which determine the current sensitivity of a moving-coil galvanometer and explain how they are interrelated.
(b) Distinguish between the sensitivity and accuracy of a galvanometer and suggest what factors influence accuracy.
(c) Why is there a practical limit to the sensitivity of a galvanometer?
(d) Why can a suspended-coil type of galvanometer have a greater sensitivity than a pivoted type? What are the advantages of a taut suspension meter? ■

## Q 3.30 Self-assessment question

A coil of 100 turns is wound on a rectangular frame sides $1.2 \times 10^{-2}$ m and $2.0 \times 10^{-2}$ m. The coil is pivoted and swings in a radial field of $1.5 \times 10^{-2}$ T. The torsion constant $c$ of the hairspring is $1.8 \times 10^{-6}$ N m rad$^{-1}$. A current passing through the coil causes a steady deflection of 0.50 rad.

Calculate

(a) the magnetic deflecting torque (in terms of $I$),
(b) the elastic restoring torque,
(c) the magnitude of the current,
(d) the current sensitivity (in radians per mA),
(e) the current sensitivity in mm per mA, if the length of the pointer is 40 mm. ■

## Ammeters and voltmeters

A moving - coil galvanometer is an instrument which detects small currents passing through it. With a calibrated scale, it becomes a milliammeter or microammeter. Resistors connected in parallel (shunts) or series (multipliers) can be used to convert the meter to measure larger currents (ammeter) or potential difference (voltmeter). This is considered in the unit *Electronic systems*.

*Note.* The majority of ammeters now are digital ammeters.

## Electric motors

An electric motor is a device which works by passing an electric current through a coil in a magnetic field (figure 3.20a). The problem with a simple coil is that it will turn until it is perpendicular to the $B$-field and then come to rest (figure 3.20b). To continue the motion, the current needs to be reversed so that the forces are reversed and the coil continues to turn.

The coil will continue to turn because it always slightly overshoots the vertical position (figure 3.20c), and so the reversed force can maintain the motion.

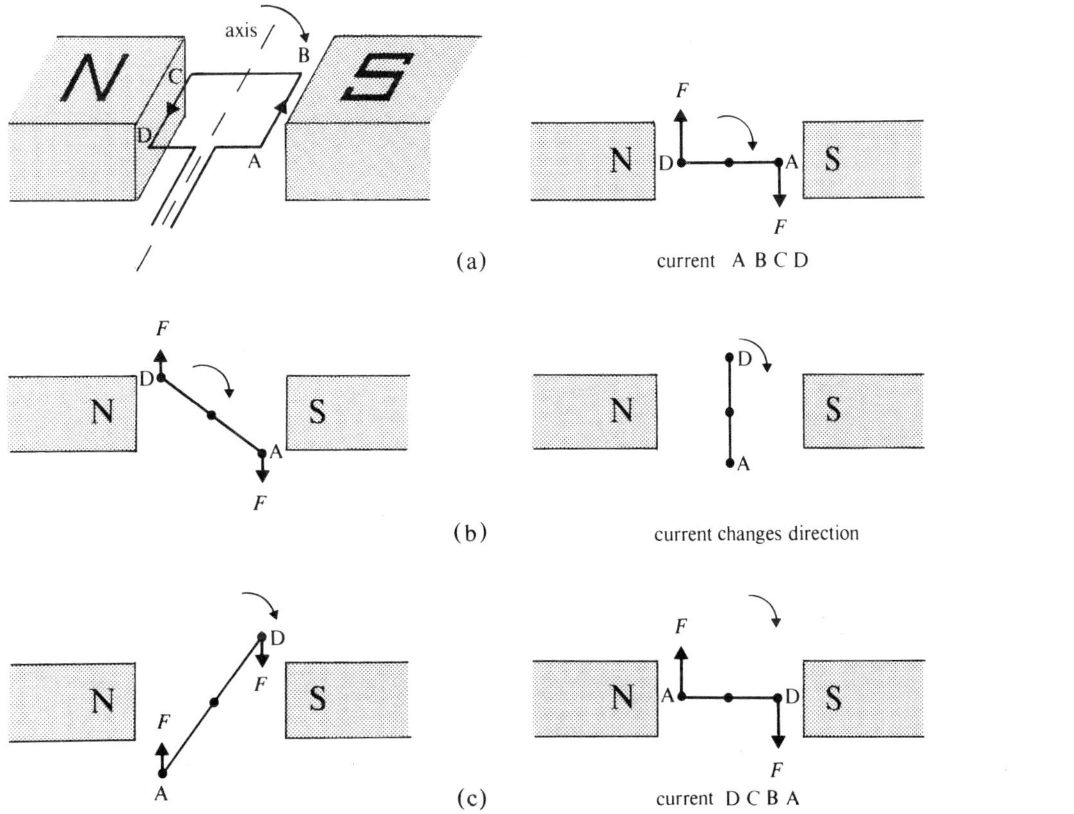


**Figure 3.20**

The device for reversing the current is known as the **commutator**.

### Q 3.31 Study question

(a) Describe with the aid of labelled diagrams, the construction and operation of a simple d.c. electric motor with a commutator.

Explain the action of a commutator in maintaining the motion of the coil.

(b) Examine a normal commercial motor. What are the principal differences between this and a simple d.c. motor?

(c) Explain how these differences make the commercial motor more powerful. ■

EXPERIMENT FF5

# Force on a current in a magnetic field

## Apparatus

- ❍ l.t. variable voltage supply or 3 accumulators or car battery
- ❍ rheostat 0-15 Ω, 5 A
- ❍ 2 ammeters 0-5 A
- ❍ plotting compass
- ❍ wide current balance
- ❍ 3 pairs of magnadur magnets with steel yokes
- ❍ high inductance coil (1100 turns)
- ❍ flat solenoid
- ❍ thin wire or paper for counterweight (rider)
- ❍ 1 m copper wire s.w.g. 14
- ❍ connecting leads

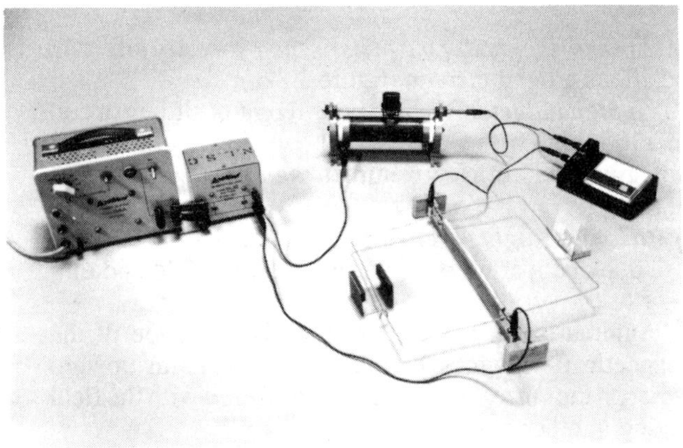

**Figure 3.21**

A convenient arrangement for studying the magnetic force on a current is shown in figure 3.21. The copper wire frame is balanced on metal edges, through which a current is supplied to one arm. Current does not flow in the other arm because there is an insulating section. This usually has a pin through it to stop the balance tilting too far. If the arm carrying the current is placed in a magnetic field it will swing if the force is upwards or downwards. To compare the magnitude of forces, the magnetic force can be made to act upwards. It can then be balanced by the gravitational force on a counterweight. The balance becomes more stable if it is bent slightly about the pivots, but note that the more it is bent the less sensitive it becomes.

You will need to provide power supplies for two circuits and be able to vary the currents (i) in the current balance (ii) in a straight wire, coil or solenoid placed near the balance.

### Direction of force

1 Set up the current balance, and connect it to the supply circuit (figure 3.21). Adjust the current in the balance arm to between 3 A and 5 A. (Switch off the currrent when you are not making an observation: it can cause heating at the pivots, and they can weld themselves to the wire!)

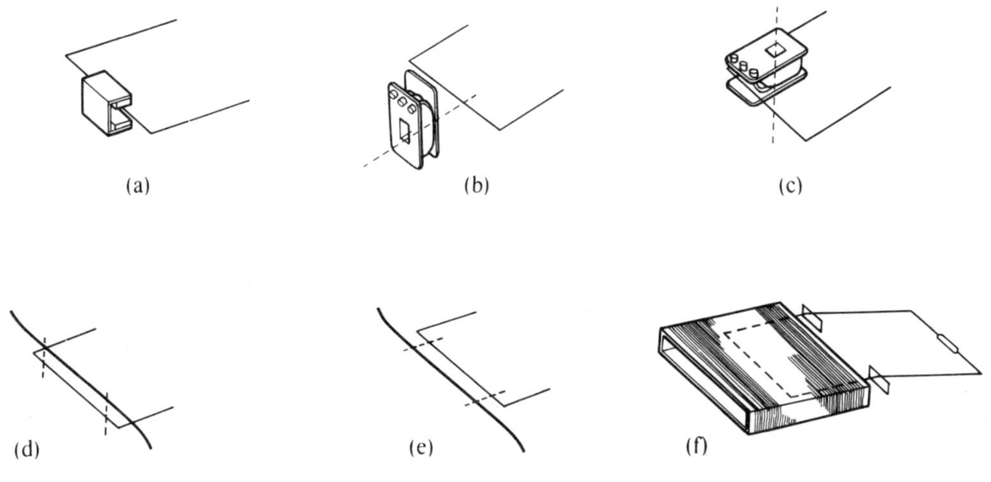


**Figure 3.22**

continued

## EXPERIMENT FF5

2 Use the balance to find the direction of the force due to the fields of

(a) *a pair of magnadur magnets on their yoke*: try different positions around the arm (figure 3.22a);
(b) *a high inductance coil*: try different positions (a coil of 1100 turns with a current of 1-2 A flowing through it will give an effect a few centimetres away) (figures 3.22b and c);
(c) *a long straight wire*: a high current will be needed, so use thick copper wire (14 s.w.g) (figures 3.22d and e).

3 Summarise your results stating the direction of the magnetic field which produces the maximum upward force on the current and the effect produced with the field acting in other directions.

**Factors affecting the size of the force**

1 Place one pair of magnadur magnets (on their yoke) around the current-carrying arm of the balance, so that it is forced upwards. Make a small counterweight (of wire or paper) and place it on the arm of the balance which passes through the magnetic field.

2 Use a counterweight which will be suitable for a current between 1 and 2 A. Adjust the current until the balance just swings over and note this current value.

3 Add a second equal counterweight; adjust the current until the balance just swings over and record the value of the current. Repeat with a third counterweight.

4 Is the force proportional to the current flowing through the field?
*Note*. The force clearly increases as current increases and the result $F \propto I$ is actually inevitable with the meter you are using! The meter works because a force acts on the current passing through a magnetic field inside the meter and the scale is calibrated to show double the current when this force doubles. But it is still a convincing exercise.

5 Adjust the current so that the balance just swings over (and the pin moves down) using one counterweight and one pair of magnets.

6 Place a second pair of magnets alongside the first so that the length of current on which the field acts is doubled. Keep the current constant but add counterweights to counterpoise the balance. Repeat the procedure with a third pair of magnets.

7 Investigate the effect of varying the magnetic field. The current in the balance must be kept constant and the arm of the current balance must be placed close to one end of the 1100 turn coil or inside the flat solenoid (see figure 3.22f). The magnetic field can be altered by varying the current in the coil or solenoid and you can assume that the field strength is proportional to the current in the coils.

8 Write a summary of your results stating how the force on the current depends on the factors you have been investigating.

EXPERIMENT FF6

# Investigation of magnetic flux density using a Hall probe

## Apparatus

- ○ Hall probe, circuit box and dry cell
- ○ picoammeter or digital ammeter
- ○ set of solenoids
- ○ magnetic field board
- ○ d.c.power supply 0-12 V, 8 A
- ○ ammeter 0-10 A d.c.
- ○ rheostat 0-15 Ω,5A
- ○ 2 slotted bases or G clamps

### Solenoids

1 Set up the circuit as shown in figure 3.23. Adjust the rheostat so that there is a current of about 1.0 A through the solenoid. (Arrange for the power supply to be distant from the solenoid so that it does not contribute to the magnetic field.)

Connect the Hall probe and the circuit box to the sensitive meter. Adjust the zero control, if necessary, so that there is no deflection when the probe has no field passing through it.

2 Place the probe well inside the solenoid with the slice perpendicular to the axis of the solenoid. Adjust the sensitivity of the meter to obtain a large deflection. (If you change the sensitivity during an investigation, record this fact and multiply the succeeding deflection by the appropriate factor in your results table.)

3 Now place the probe so that the centre of the slice coincides with the centre of the axis of the solenoid and record the meter deflection. Move the probe along the axis away from the centre and record probe position and meter deflection at intervals until the probe is a short distance outside one end. Record your results graphically and so find (i) over what fraction of the length of the solenoid the field is uniform, and (ii) the ratio of the flux densities at the centre of an end face and right inside the solenoid.

4 The direction of the field at a point is perpendicular to the slice when the meter is showing maximum deflection for that point. Use the probe to find how the field varies in magnitude and direction over an end face. Is there a field outside and alongside the solenoid?

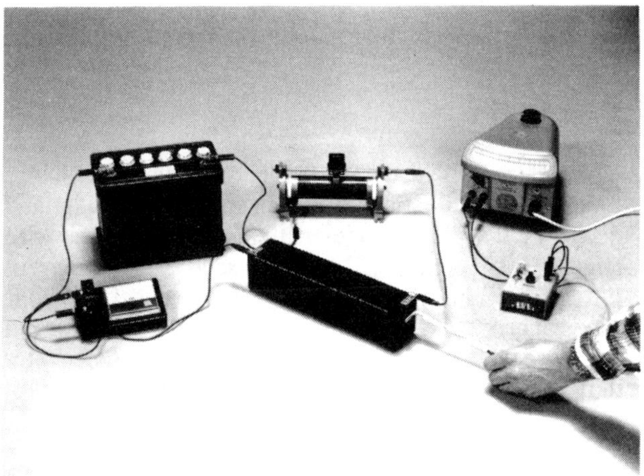

**Figure 3.23**

5 Having investigated the variation of field near a solenoid you can now find what factors affect the flux density at the centre of a long solenoid. By varying the current in a solenoid and then by sending the same current through several solenoids (a way of doing this is to connect them in series) find the answers to the following questions.
How does $B$ depend on $I$? (Give a set of results or a graph to support your answer.)

Does the cross-sectional area affect the strength of the field? (Use two solenoids of equal length and density of turns but different cross-sections.) How does flux density depend on this number of turns per unit length?

continued

**EXPERIMENT FF6**

## Circular coil

1 Measure the flux density at the centres of several flat coils each with equal numbers of turns but with different radii. The coils can be laid out on the magnetic field board (figure 3.24) and the same current (of several amperes) sent through each.

2 How does the flux density depend on the radius?

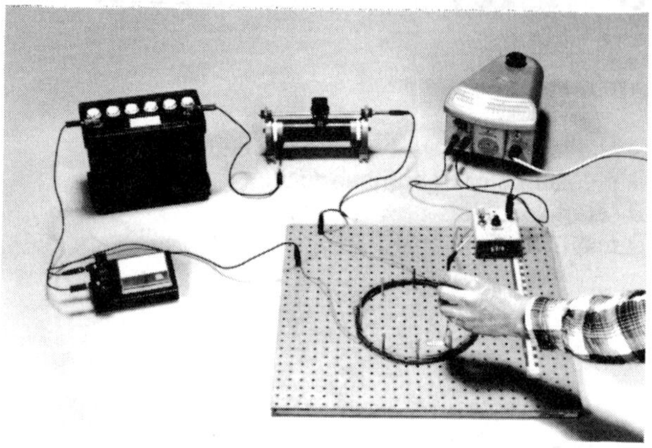

**Figure 3.24**

## Straight wire

You can investigate the field around a straight conductor to test whether the field is at right angles to the wire and note how the field changes with distance from the wire (figure 3.25). You will need to send a large current through the wire, up to 10 A, but you must check that your rheostat is rated to take the current you are providing and switch on only when you are taking a reading of the meter deflection.

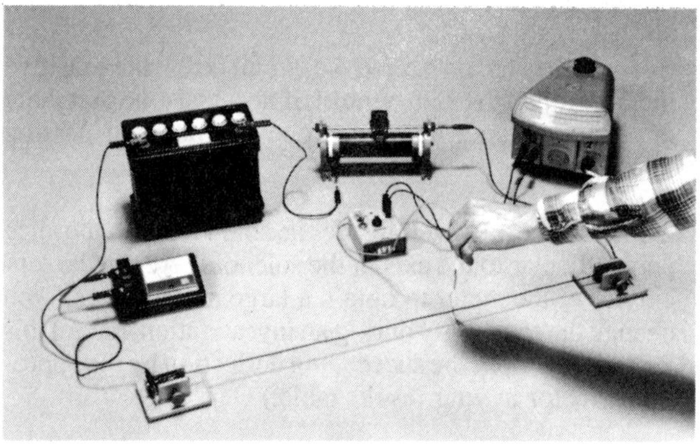

**Figure 3.25**

# Questions on objectives

1 Figure 3.26 shows patterns of magnetic field lines in the vicinity of magnets and current-carrying conductors. For each arrangement, write a sentence or two to describe what you can deduce from the magnetic field patterns.
*(objective 1)*

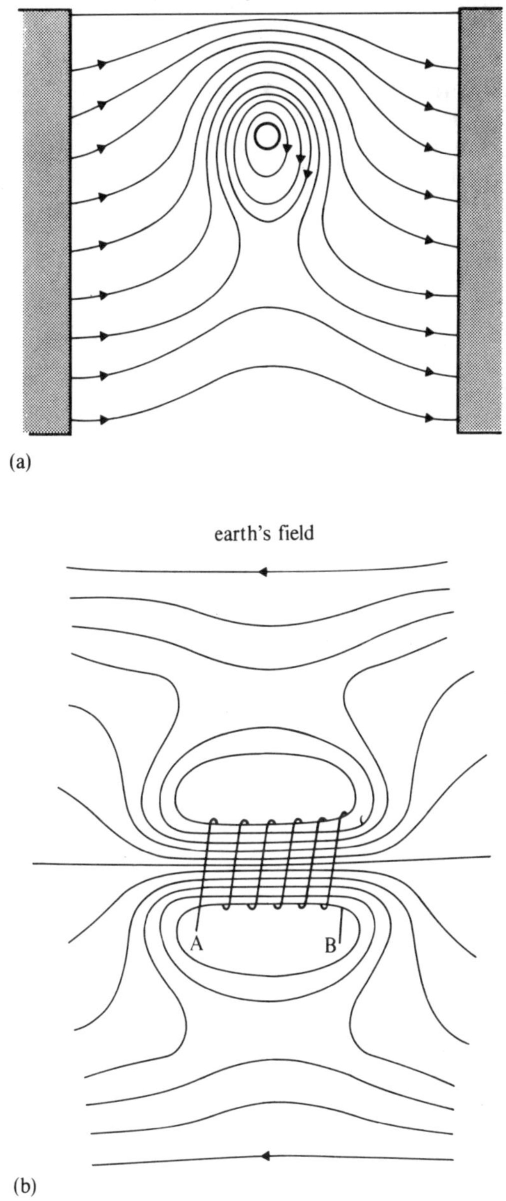


**Figure 3.26**

2 Figure 3.27 shows arrangements of current-carrying conductors. Answer the questions below and explain your answers briefly.

(a) Which way would wire B move in figure 3.27a?
(b) What would happen to the coil shown in figure 3.27b?
(c) In figure 3.27c which way would the straight wire move?
*(objective 6)*

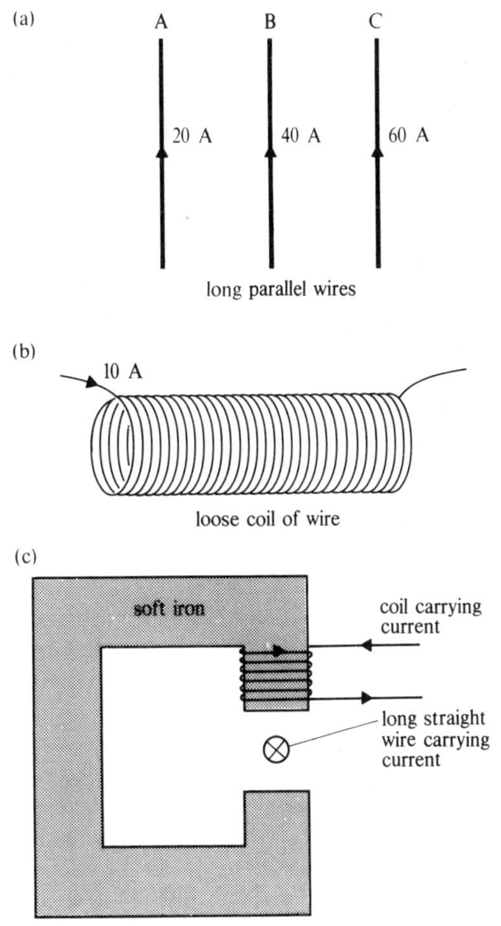


**Figure 3.27**

3 (a) Define the ampere.
(b) What are the problems of using this definition directly to measure the ampere?
(c) Explain briefly how the measurement of electric current by a moving-coil ammeter is related to the definition, but overcomes the problems in (b). *(objectives 2 and 10)*

4 Describe briefly how you would measure the magnetic flux density between the poles of a horseshoe magnet. State the principles on which your method depends. *(objective 7)*

5 A straight conductor of length 0.20 m carries a current of 1.5 A, in a magnetic field of flux density 80 m T. What is the force which acts upon the conductor when it is (i) perpendicular to the field, (ii) inclined at 30° to it? *(objective 8)*

6 Two long cables providing an electricity supply run horizontally 2.0 m apart and each carries a current of 10 A in a direction west to east. The earth's magnetic field has a flux density of $5.5 \times 10^{-5}$ T and is directed northwards and at 70° to the horizontal.
(a) Draw a sketch of the cables showing the direction of the magnetic force acting on each cable.
(b) Calculate the magnitude of these forces per unit length of cable.
(c) What other forces act on each element of the cable?

Which of these forces change in magnitude *and* direction when the current is switched on? ( $\mu_o = 4\pi \times 10^{-7}$ N $A^{-2}$)
*(objectives 4, 6 and 8)*

7 The current $I$ through a slab of $n$-type germanium is caused by a drift of electrons. If the electrons move with average velocity $v$ through a cross - sectional area $A$, $I = nAev$ where $n$ is the number of conduction electrons per unit volume each carrying charge $e$. Use this equation to derive an expression for the Hall p.d. produced when a magnetic field is applied across the slab.

What flux density applied perpendicular to a thin slice of $n$-type germanium 1.0 mm thick, will produce a Hall p.d. of 10 mV if the current is 50 mA and the number density of conduction electrons is $6.0 \times 10^{20}$ $m^{-3}$? (charge on electron = $-1.6 \times 10^{-19}$ C) *(objective 9)*

8 An air-cored solenoid is 1.2 m long, with 3000 turns of diameter 80 mm. When it carries a current of 3.0 A calculate the magnetic flux density along the axis

(a) at the centre,
(b) at one end. ( $\mu_o = 4\pi \times 10^{-7}$ N $A^{-2}$)
*(objective 8)*

9 A commercial electric motor has the following features: (i) a soft iron core, (ii) several coils, (iii) an electromagnet for the field coils, (iv) a commutator with many splits.

Explain why each of these is necessary and how they make the motor more powerful. *(objective 10)*

10 A suspended - coil galvanometer has a coil of 60 turns of area $5.0 \times 10^{-4}$ $m^2$ which is suspended by a bronze strip of torsion constant 0.30 N $rad^{-1}$. When a current of 25 A is passed through the coil a spot of light, reflected from a mirror attached rigidly to the coil, is deflected 30 mm on a screen which is 1.0 m away. What is the flux density in the radial field in which the coil is suspended? *(objective 12)*

TOPIC 4

# Motion in fields

**Study time 1 week**

## Summary

In this topic you will apply the models that have been developed about fields to a study of the paths of bodies moving in gravitational fields and of charged particles moving in electric and magnetic fields. The photograph shows a proton accelerator.

## Objectives

When you have completed the work in this topic you should be able to:

1 Use the following scientific terms correctly: escape velocity, synchronous satellite.

2 Derive an expression for the speed of a body which is moving in a circle in a gravitational field.

3 Solve problems on the motion of a body in a uniform gravitational field.

4 Describe, and draw diagrams to illustrate, how charged particles are deflected by uniform electric and magnetic fields.

5 Show that if a charged particle moving with constant velocity enters a magnetic field which is at right angles to its direction of motion it will move in a circular path.

6 Solve problems on the motion of charged particles in electric and magnetic fields.

7 **EXTENSION**

Derive, and use, an expression for the escape velocity of a body from the gravitational field of the earth or a planet.

8 **EXTENSION**

Show that if a charged particle moving with constant velocity enters an electric field which is at right angles to its motion, it will move in a parabolic path.

## References


| | |
|---|---|
| Akrill | Chapter 12 |
| Bennet | Chapters 12 and 14 |
| Duncan | Chapters 8 and 22 |
| Muncaster | Chapter 50 |
| Nelkon | Chapters 2 and 31 |
| Wenham | Chapter 33 |
| Whelan | Chapters 20 and 52 |

## 4.1 Motion in a gravitational field

This section is not just about objects falling to the ground with uniform acceleration but it is concerned with the motion of planets, artificial satellites and space probes, and even the escape of gas molecules from the earth's atmosphere.

The data of table 4.1 will be useful in answering the questions in this section.

**Table 4.1**

| | |
|---|---|
| Mass of earth $M$ | $6.0 \times 10^{24}$ kg |
| Mass of moon $M_m$ | $7.0 \times 10^{22}$ kg |
| Radius of earth $R$ | $6.4 \times 10^{6}$ m |
| Distance earth to moon (i.e. average orbit radius) | $3.8 \times 10^{8}$ m = 60 $R$ |
| Gravitational constant $G$ | $6.7 \times 10^{-11}$ N m$^2$ kg$^{-2}$ |

Throwing things away is easier said than done. Things have a tendency to stay around and it is perhaps a good job that gravitational forces prevent us from spreading our litter too liberally around the whole universe. If we throw an object up we know that at its maximum height, its kinetic energy has been transformed into extra potential energy, and $mv^2 = mgh$ (figure 4.1a).

This, of course, is only true if the gravitational field strength $g$ is uniform. What happens when an object is projected to a greater height so that we cannot assume uniformity of field? The gravitational field around the earth is an inverse square law field - the field which would be produced if all the mass of the earth were concentrated at its centre (figure 4.1b).

The gravitational potential at any point distance $r$ from the earth's centre is

$$\frac{-GM}{r} \quad (\text{for } r > R)$$

The change in gravitational potential between two points is numerically equal to the work done in moving one kilogram between these points.

### Q 4.1 Development question*

(a) Write down an expression for the change in potential energy per kilogram which occurs when a body is moved from A to B (figure 4.1b) assuming negligible air resistance.
(b) How much kinetic energy must a body of mass $m$ have when it is projected from A (i) if it is to rise to B, (ii) if it is to escape from the earth's gravitational field?
(c) If a body leaves the earth at speed $v_e$ show that it will just escape from the earth's gravitational field if

$$v_e = \sqrt{\left(\frac{2GM}{R}\right)}$$

(d) Use the data provided in table 4.1 to calculate (i) the escape speed $v_e$ from the earth and (ii) the escape speed from the moon. ■

Notice that in the last question speed $v_e$ was calculated by considering energy changes and there was no need to specify a direction. A body with this speed will escape if it is projected vertically or horizontally or in any direction above the horizontal. If a body is projected with speed $v$, it will travel in a parabolic path (unless it is projected vertically) - see figure 4.2.

Man-made rockets are not the first bodies to have escaped from the surface of the earth or moon. Gas molecules moving with speeds which are determined by the temperature may move fast enough to escape from a planet's atmosphere. The random collision process in a gas means that there will always be a few very high speed molecules which can escape but there will be a very low probability of molecules escaping if the root-mean-square speed (r.m.s. speed) of the molecules is much smaller than the escape speed.

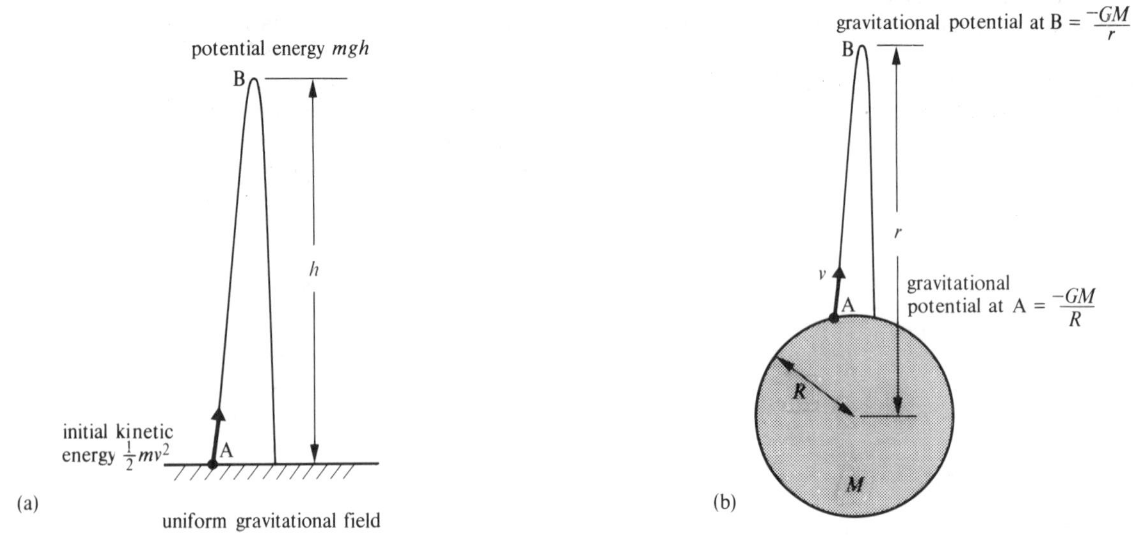


**Figure 4.1**

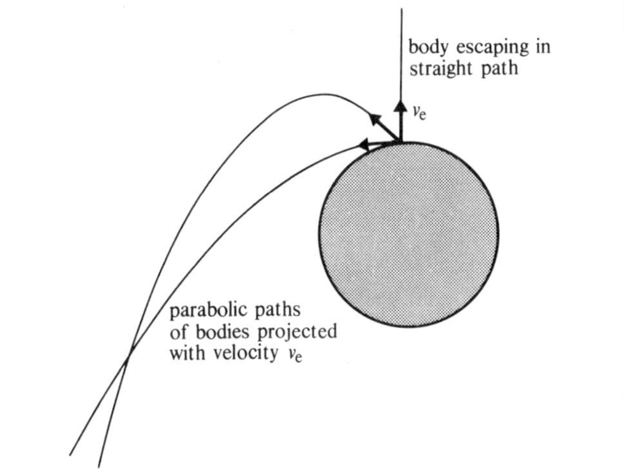


**Figure 4.2**

## Q 4.2 Self-assessment question

The r.m.s. speed of hydrogen and oxygen molecules at 300 K is 1.8 km $s^{-1}$ and 0.35 km $s^{-1}$ respectively. Use your answer to question 4.1(d) to explain

(a) why the earth's atmosphere contains no hydrogen, and
(b) why the earth has retained an atmosphere but there is no atmosphere around the moon. ■

## Satellites

Suppose a body is projected horizontally from a point, distance $r$ from the centre of the earth (where there are no atmospheric resistance forces acting).

Suppose that after projection the body moves, with a speed $v_o$, in a circle whose centre is the centre of the earth.

## Q 4.3 Development question*

(a) Why will speed $v_o$ remain constant if the body moves in a circle around the earth's centre?
(b) What is the centripetal force which acts to produce uniform circular motion?
(c) Show that the speed $v_o$ of the satellite moving in a circular orbit is given by

$$v_0 = \sqrt{\left(\frac{gM}{r}\right)}$$

(d) Express the period $T$ in terms of $v_o$ and the orbit radius $r$. Hence show that

$$T^2 = \frac{4\pi^2 r^3}{GM} \quad ■$$

## Q 4.4 Self-assessment question

A satellite is travelling in a circular orbit at a height of 200 km above the earth's surface. ■

(a) Use the data provided in table 4.2 to calculate
(i) its orbital speed, (ii) the time for one revolution around the earth.

(b) Comment on any data provided in table 4.2 which supports your answer. ■

**Table 4.2**

| Name | Launch date | Period/mins | Height/km |
|---|---|---|---|
| Sputnik 1 | Oct. 1957 | 96.2 | 215 - 239 |
| Vostok 1 | April 1961 | 89.3 | 169 - 315 |
| Mercury 6 | Feb. 1962 | 88.6 | 159 - 265 |
| Early Bird | April 1965 | 1437 ( 24 hours) | 35 000 |

The earth's oldest satellite is the moon and measurements on its orbit provide support for the inverse square law of gravitation.

## Q 4.5 Development question*

In this question you can assume that the radius of the earth $R$ is 6400 km and that the gravitational field strength $g$ at the surface, distance $R$ from the centre, is 10 N $kg^{-1}$.

(a) The moon is orbiting at a distance $60R$ from the earth's centre. What is the force on each kilogram of the moon due to the earth's gravitational field?
(b) If the force provides the necessary centripetal force to maintain the moon in a circular orbit calculate the moon's orbital speed.
(c) Hence, calculate the orbit period for the moon (i) in seconds (ii) in days (giving 2 significant figures). ■

How can we 'weigh' the earth? One way is to use the known orbit time and orbit radius of the moon, together with the key experimental value for $G$.

## Q 4.6 Self-assessment question

Using a value for the radius of the moon's orbit and for $G$ (table 4.1) and taking the orbit period for the moon as 27.7 days, obtain a value for the mass of the earth. ■

One of the most important uses of satellites is in world-wide communications. Communication satellites are 'synchronous' satellites, so called because their orbit period is the same as the period of rotation of the earth - 24 hours. They also travel in the same direction as the earth rotates and this means they are always above the same place on the earth's surface - a point somewhere on the equator. (Three such satellites are needed, one over each of the main oceans, Pacific, Indian and Atlantic, to provide a world-wide communication network.) Radio waves can be beamed to these satellites, and they amplify the received signal and retransmit it to another part of the earth.

The relationship $T^2 \propto r^3$ applied to the sun's satellites (the planets) is a statement of Kepler's third law. The relationship is also true for the earth's satellites (see question 4.3d) and you can now use it to calculate the height of a communication satellite.

## Q 4.7 Self-assessment question

(a) If $R$ represents the radius of the earth then the moon is distance $60R$ from the earth's centre, and has an orbit period

of 27.7 days. If the orbit radius of a synchronous communication satellite is $xR$ calculate the value of $x$ assuming $T^2 \propto r^3$.
(b) If $R$ = 6400 km, what is the height *above the earth* of a communication satellite in circular orbit? Does any data given in table 4.2 confirm your answer? ■

### Q 4.8 Self-assessment question

(a) Use the data in table 4.1 to show that at a distance from the moon which is about 10 % of the total earth-moon distance there is a gravitational neutral point.
(b) Figure 4.3 shows the speed of the Apollo 11 spacecraft during its return journey to the earth. During the part of the flight recorded on the graph, the rocket engines were switched on once only for a short mid-course correction. Explain the shape of the graph around A, B and C. Does the graph give support for your answer in part (a)? ■

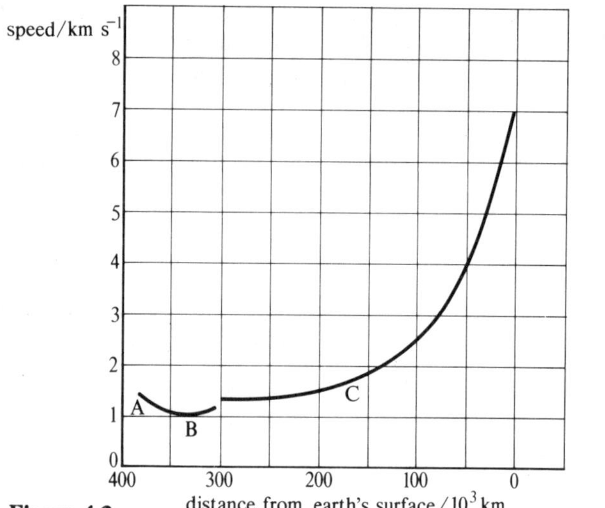


**Figure 4.3**

## 4.2 Motion in electric and magnetic fields

In this final section of the unit you will study the motion of charged particles in electric and magnetic fields. You will consider how electrons can be *accelerated*, *deflected*, and *focused* by these fields, and this knowledge will be applied to understand the principles of the oscilloscope, the electron-microscope and the cyclotron. This section is also an important preparation for work in the unit *Electrons and the nucleus* when you will consider the mass spectrometer, X-ray machines, and the deflection of high speed ions in the field around a nucleus.

### Accelerating fields

In instruments in which electron streams are deflected by electric and magnetic fields an electron gun is the source of high-speed electrons. Electrons liberated from a heated metal oxide surface are accelerated by an electric field between cathode and anode (figure 4.4).

A force is exerted on the electron by the electric field and work is done on it. The electrons gain kinetic energy.
Some electrons will pass through the hole in the anode to produce a beam of electrons travelling at a constant speed in a region where there is no accelerating field.

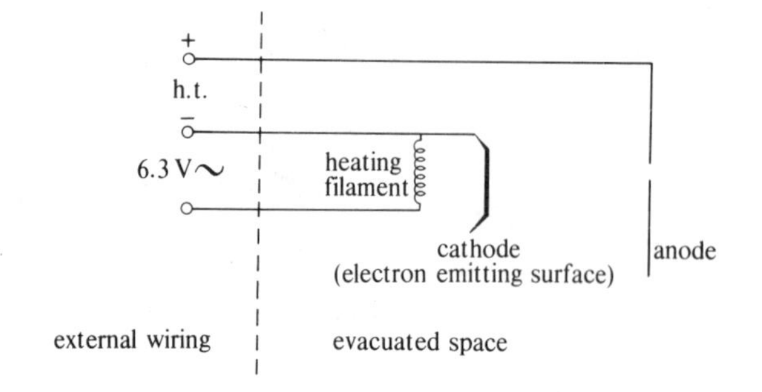


**Figure 4.4**

### Q 4.9 Development question*

If the p.d. between the cathode and anode is $V$ show that the speed of an electron of mass $m_e$ is given by

$$v = \sqrt{\left(\frac{2eV}{m_e}\right)}$$

where $e$ is the magnitude of the charge on the electron.

*Note.* The speed does not depend on the separation of the electrodes. ■

### Q 4.10 Self-assessment question

A potential difference of 500 V is applied between the anode and cathode, in an evacuated tube, of an electron gun. Calculate

(a) the electric field between the anode and cathode if they are separated by a distance of 5.0 cm,
(b) the force acting on an electron in the field between anode and cathode,
(c) the kinetic energy acquired by the electron in moving between anode and cathode,
(d) the speed at which an electron will emerge from the gun assuming that it was emitted from the cathode with a negligible small speed,
(e) the speed at which electrons will emerge from the gun if the separation between anode and cathode is reduced to 2.5 cm.
(magnitude of charge on the electron = $1.6 \times 10^{-19}$ C, mass of electron = $9.1 \times 10^{-31}$ kg) ■

### Deflection in an electric field

### Q 4.11 Self-assessment question

If a fine beam of electrons passes through the electric field between charged horizontal plates P and Q (figure 4.5), the path of the electrons is a **parabola** - the sort of path that a stone moves in when thrown horizontally near the earth's surface.

(a) What does this tell you about the potential of Q compared to that of P?
(b) What does this suggest about the horizontal motion of the electrons?
(c) What can you say about the vertical force and acceleration of the electrons?
(d) What is the path of the electrons beyond the plates?

(e) If there is no potential difference between P and Q, will the path of the electrons be a parabola? Give a reason for your answer.
(f) Consider typical values for the deflecting field (e.g. 500 V between plates 5 cm apart), and explain by an order-of-magnitude calculation why gravitational forces can be ignored. ■

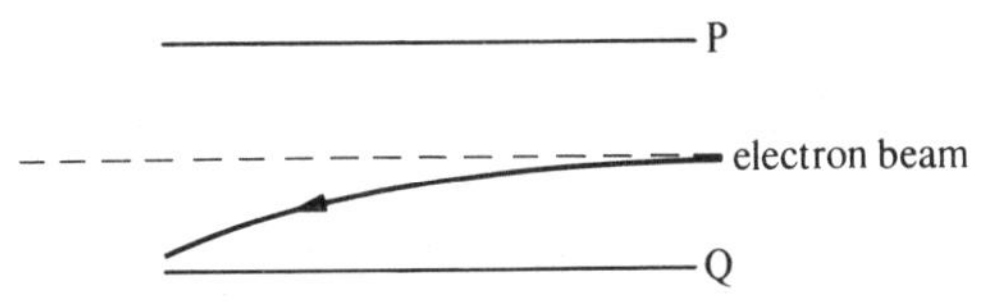


**Figure 4.5**

## EXTENSION

### Q 4.12 Development question*

This question will help you to derive the equation for the parabolic path taken by an electron which enters an electric field $E$ with an initial speed $v$ perpendicular to the field (figure 4.6).

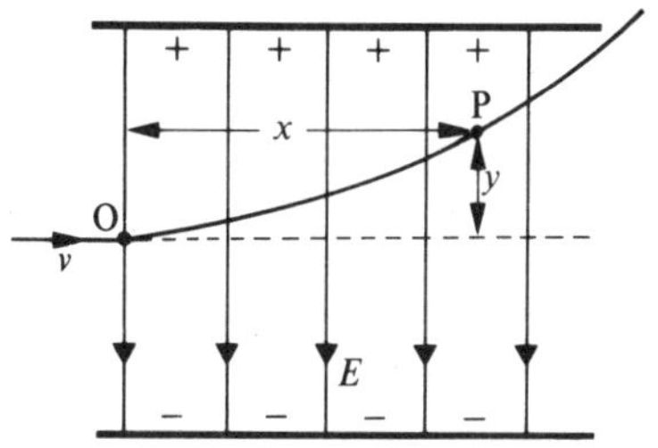


**Figure 4.6**

P is the position of the electron at time $t$ after passing through O.

(a) Write an equation for $x$ in terms of $v$ and $t$.
(b) Derive an expression for the acceleration of the electron due to field $E$, and hence write an equation for the displacement $y$ after time $t$.
(c) Use the equation derived above to show that

$$y = \frac{1}{2}\left(\frac{Ee}{m_e v^2}\right)x^2$$

(d) The general equation for a parabola whose axis is the $y$ axis is $y = kx^2$, where $k$ is a constant. What are the conditions for the electron's path to be a parabola? ■

## Deflection in a magnetic field

When an electron enters a uniform magnetic field and moves at right angles to the field it experiences a force $F_B$ which acts in a direction which is perpendicular to that of this field and perpendicular to the direction of motion (figure 4.7). This force will cause the electron to move in a **circular** path.

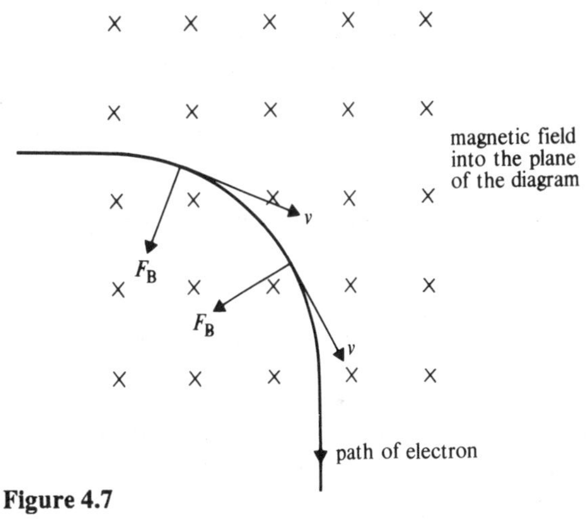


**Figure 4.7**

### Q 4.13 Development question*

(a) Why is the speed of the electron constant?
(b) Why is the magnitude of the magnetic force on the electron constant?
(c) Derive an expression for the radius of the electron's orbit in terms of $B$, $e$, $m_e$ and $v$.
(d) Using the expression for the speed of an electron accelerated by an electric field $e/m_e$ (obtained in question 4.9), derive an expression for $e/m_e$ in terms of $B$, $v$ and $r$. ■

### Q 4.14 Self-assessment question

Calculate the magnetic flux density which is required to make an electron travelling with a speed of $8.0 \times 10^6$ m s$^{-1}$ move in a circular path of radius $5.0 \times 10^{-2}$ m. (The ratio of charge to mass for an electron, $e/m_e$, is $-1.76 \times 10^{11}$ C kg$^{-1}$) ■

If the velocity of the electron has a component *along* the direction of the magnetic field then the electron will travel in a helical path (figure 4.8a).

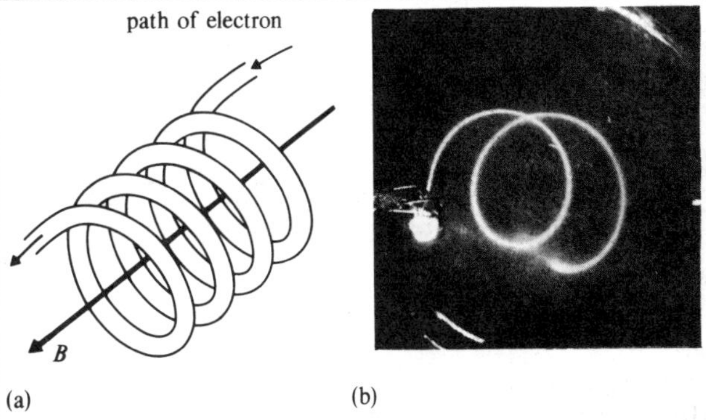


**Figure 4.8**

Figure 4.8b is a photograph of the helical path of electrons in a fine beam tube. As the high speed electrons travel in this path they ionise residual gas atoms in the tube which contains low pressure hydrogen. The gas ions emit light where they recombine with electrons and so provide a means of tracking the path of the electrons.

## Q 4.15 Development question*

A beam of electrons moving with uniform velocity $v$ enters a uniform electric field of field strength $E$ at right angles to their direction of motion. A magnetic field of flux density $B$ is applied at right angles to the electric field.

(a) Show clearly on a diagram the direction in which the electric and magnetic fields must be applied so that the forces exerted on an electron in the beam are in opposite directions.
(b) Write down expressions for the forces exerted on an electron.
(c) Assuming that $B = 1.0 \times 10^{-2}$ T and $E = 100$ V m$^{-1}$, calculate a value for the speed of the electrons if the beam passes through the fields without being deflected. ■

Knowledge of nuclear structure can be obtained by collision experiments in which high speed particles are used to bombard matter. You have seen how charged particles can be accelerated by an electric field using a high potential difference between electrodes. In the cyclotron, accelerating fields are used in conjunction with a deflecting magnetic field to obtain a beam of high speed positively charged particles. These are produced near the centre of an evacuated box formed by two D-shaped electrodes (figure 4.9).

A uniform magnetic field provides the centripetal force to enable a particle of charge $Q$ and mass $m$, to move in a circular orbit of radius $r$, perpendicular to the magnetic field.

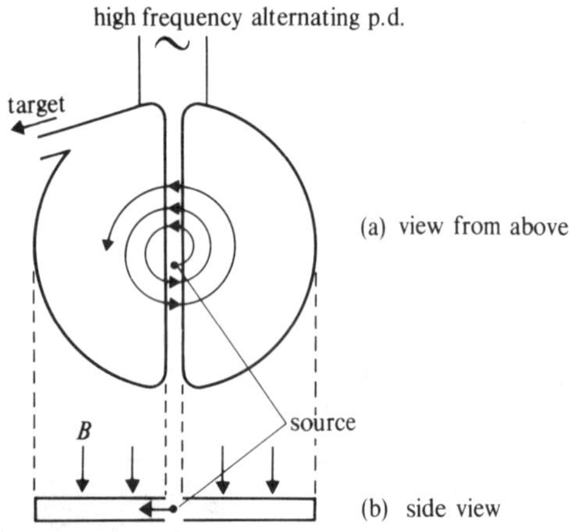


**Figure 4.9**

## Q 4.16 Development question*

(a) What happens to the orbit radius as the speed of the particles increase?
(b) Derive an expression for the period of revolution $T$ of the particles in circular orbit.
(c) What happens to the period $T$ as $v$ increases?

The electrodes in a cyclotron must be connected to a supply of alternating p.d. so that the particles always pass through an accelerating field when they cross the gap between the D-electrodes.

(d) What frequency of supply is required to accelerate protons to high energy if the applied magnetic field has flux density 0.80 T?
(proton mass = $1.7 \times 10^{-27}$ kg,
proton charge = $1.6 \times 10^{-19}$ C) ■

**EXTENSION**

## Focusing fields

Oscilloscopes and television sets can function effectively only because electron beams can be focused by electric or magnetic fields. Figure 4.10 represents a section through two cylindrical electrodes. The electric field between the cylinders is represented by equipotentials and this field acts as an electron lens.

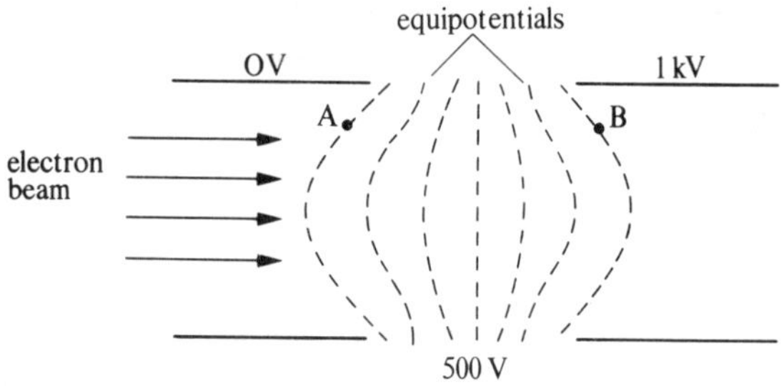


**Figure 4.10**

## Q 4.17 Development question*

(a) What effect has the field on the direction and magnitude of the velocity of an electron (i) passing through A (ii) passing through B?
(b) What will happen to the beam in the region (i) to the left of the central equipotential and (ii) to the right of the centre?
(c) How will the effect differ for high speed and low speed electrons? ■

Because the electrons are accelerated they will spend less time in the region beyond the central equipotential and the divergence produced in this region will be less than the convergence produced in the region near A. The net result is a converging effect. The field is acting as a converging electron lens.

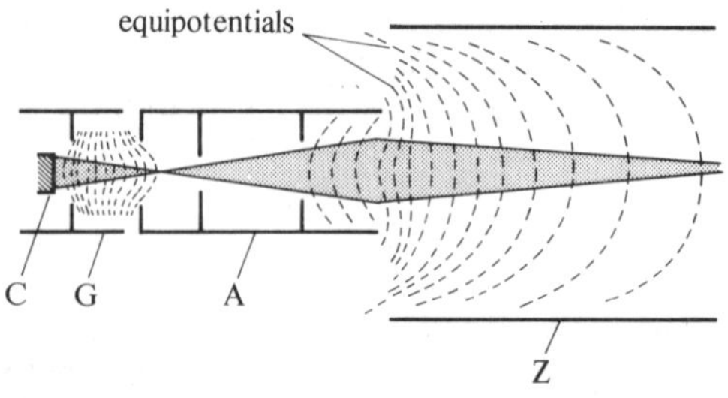


C cathode, emitting electrons
G grid controlling intensity of beam
A accelerating anode

Electron lens formed between anodes A and Z to focus the beam

**Figure 4.11**

continued

This kind of lens is used in a cathode ray tube (figure 4.11) for accelerating and focusing the electron beam. The 'focal length' can be changed by altering the potentials of the cylinders.

A transmission electron microscope is the counterpart of the optical compound microscope. A beam of electrons from an electron gun is focused onto a thin specimen (thickness of the order of $10^{-7}$ m). The transmitted electron beam is passed through magnetic lenses which act as the objective and projector lenses (figure 4.12).

A highly magnified image is projected onto a fluorescent screen to reveal the detailed structures of the specimens. Figure 4.13 shows the field around a magnetic lens formed by a circular coil wound on an iron ring.
Remember that the magnetic force on a charged particle depends on its velocity $v$ and on the component of the field which is perpendicular to $v$ ($F = Bev \sin\theta$).

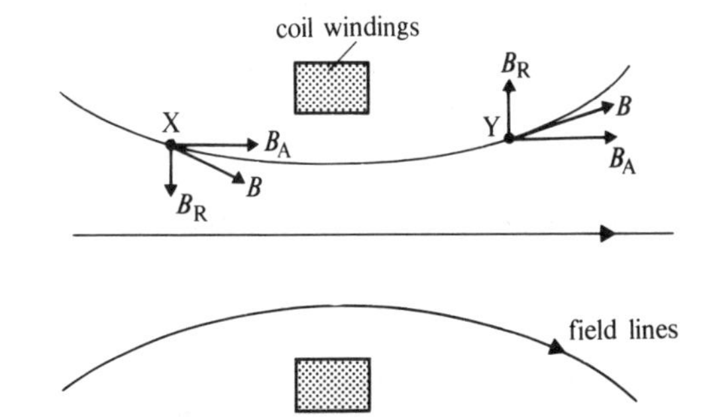


**Figure 4.13**

If an electron is travelling from left to right at X, the axial component $B_A$ of the magnetic field will have no effect but the radial component $B_R$ will exert a force out of the plane of the paper and the electron will move in a spiral round the axis of the coil.

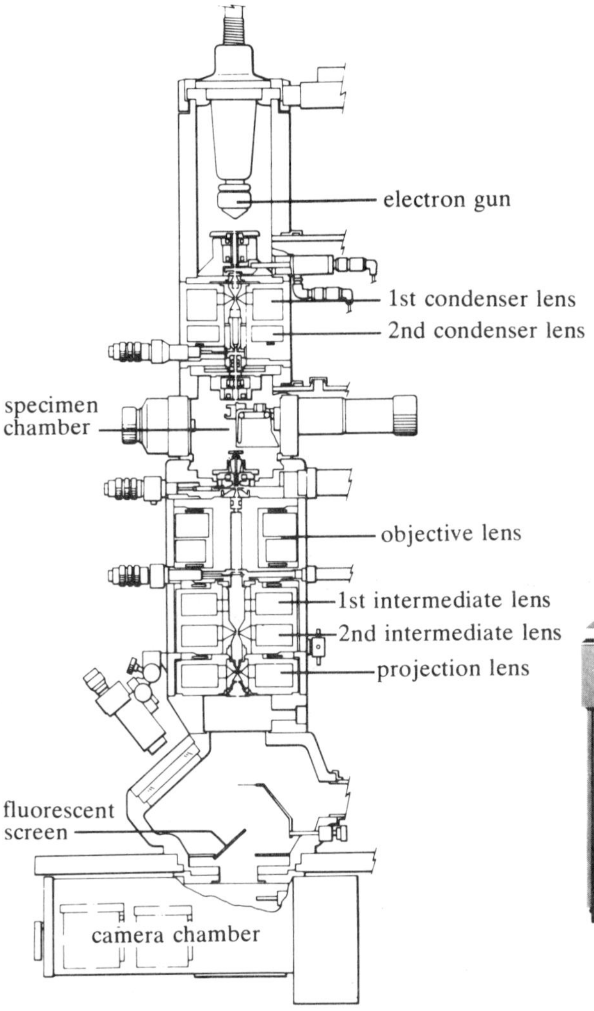


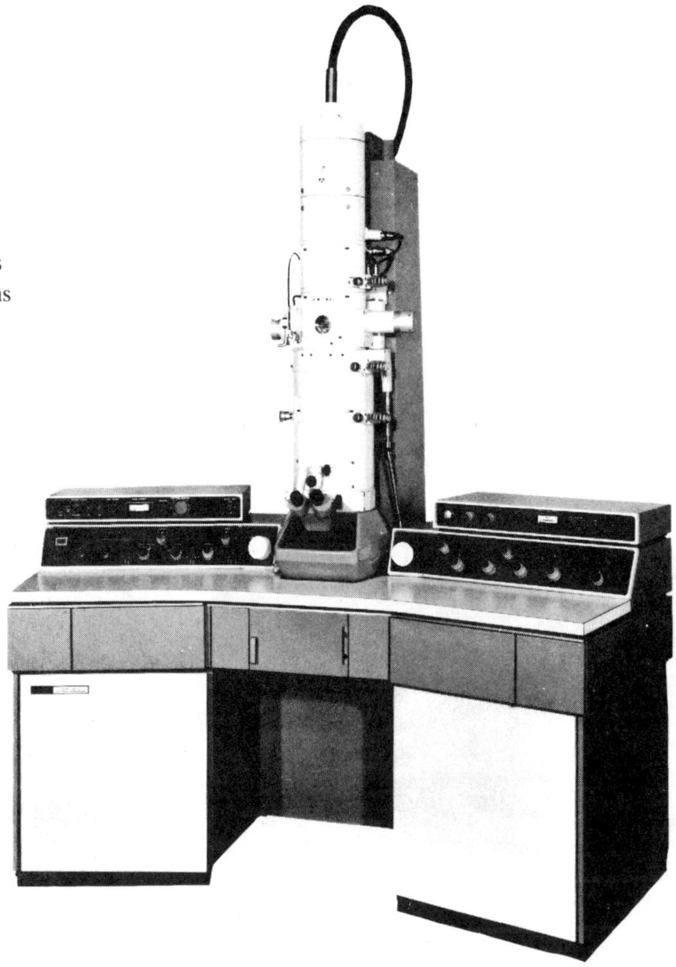

**Figure 4.12**

continued

Because the electron now has a component of velocity perpendicular to $B_A$ (out of the paper) the electron will experience a force towards the axis due to $B_A$. (Check this statement using the left-hand rule.) This force causes the beam to converge toward the axis. At a point Y a force will act due to component $B_R$ which will reduce the circular motion and the electron will spiral at a known rate but will not reverse its direction. Hence, field $B_A$ still produces a converging effect on the beam.

It is possible in this unit to make only a brief reference to some of the many important technological applications. However, it will be clear that ideas about the forces acting on charged particles in electric and magnetic fields form the basis for the carefully developed technology which gives us colour television pictures and provides devices for measuring rapidly changing electrical impulses. Important research tools for looking into the detailed structure of matter, and bombarding atomic nuclei, also use electric and magnetic fields.

You will learn more about these in the unit *Electrons and the nucleus.*

**Q 4.18 Self-assessment question**

Consider two microscopes: one uses light of wavelength 500 nm and the other an electron beam accelerated through 10 000 V. The dimensions of the instruments are roughly similar so to compare their resolving powers we can compare the wavelengths.

(a) Calculate the wavelength of the electrons by using de Broglie's formula $\lambda = h/p$ where $h$ is the Planck constant ($6.63 \times 10^{-34}$ J s) and $p$ the momentum. (charge on electron $= -1.6 \times 10^{-19}$ C and mass of electron $= 9.1 \times 10^{-31}$ kg)
(b) Compare this with the wavelength of light.
(c) The field-ion microscope uses helium nuclei instead of electrons. What approximate wavelength is obtainable? How does this compare with the separation of ions in a metal crystal? ■

(See unit *Behaviour of matter*, topic 1.)

# Questions on objectives

1 At what speed must a satellite be travelling so that it will remain in a circular orbit $2.6 \times 10^6$ m above the earth's surface? (radius of the earth = $6.4 \times 10^6$ m, mass of the earth = $6.0 \times 10^{24}$ kg, $G = 6.7 \times 10^{-11}$ N m$^2$ kg$^{-2}$) *(objective 3)*

2 An electron is accelerated from rest through a potential difference of $3.5 \times 10^3$ volts. Calculate the speed that it acquires. (Mass of the electron = $9.0 \times 10^{-31}$ kg, charge of the electron = $-1.6 \times 10^{-19}$ C) *(objective 6)*

3 A positive ion travels in a circular orbit of radius 5.0 cm, at right angles to a magnetic field of flux density 0.50 T.

(a) Show clearly on a diagram the direction of the magnetic field and the direction of the force which is exerted on the positive ion.
(b) Calculate the speed of the ion. (The ratio of charge to mass for the positive ion is $6.5 \times 10^7$ C kg$^{-1}$) *(objectives 5 and 6)*

4 Show that a charged particle projected perpendicular to a uniform magnetic field of flux density $B$ describes a circular path. Derive an expression relating the radius of the path to the value of $B$. *(objective 5)*

5 (a) Describe and account for the form of the path of a charged particle which enters a uniform electric field travelling initially with constant velocity at right angles to the electric field lines.
(b) How does the path change if (i) the field strength is increased, (ii) the initial velocity of the particle is increased? *(objective 4)*

# Answers

## Topic 1

**1.1** It probably enables the platypus to detect the presence of live prey from their muscle activity.
**1.2** (a) 10 N $kg^{-1}$
(b) 0.5 N (downwards)
(c) The ball exerts an equal and opposite force on the earth (Newton's third law).
(d) The acceleration is negligible due to the large mass of the earth ($6.0 \times 10^{24}$ kg).
**1.3** (a) The weight of a body is the gravitational force exerted on it by the earth. $g$ is the same (by experiment) for all bodies, thus the weight of a body is proportional to its mass, thus we can assume that gravitational force $\propto$ mass of attracting body.
(b) This follows from Newton's third law.
**1.7** (a) The attraction between two 1 kg masses which are 1 m apart is approximately $10^{-10}$ N.
The earth's gravitational pull on a mass of 1 kg (at the earth's surface) is approximately 10 N.
Thus, the earth's gravitational pull is $10^{11}$ times greater than the attraction between these two 1 kg masses.

$$[G] = \frac{[MLT^{-2}][L^2]}{[M^2]} = [M^{-1}\,L^3\,T^{-2}]$$

*Note.* This also shows that $G$ could be expressed in $m^3\,kg^{-1}\,s^{-2}$.
**1.8** (a) This means that the value of $G$ is the same for all gravitating systems. Its value does not depend on (i) the nature of the attracting or attracted masses (it is the magnitude of the mass and not the type of atom which determines the gravitational attraction), (ii) the physical conditions, e.g. temperature, pressure, etc.
(b) Variation with time of $g$ or planetary orbit radii.
**1.9** (a) $W/4$
The gravitational force is inversely proportional to the square of the distance. If the distance is increased by a factor of 2, the force will decrease by a factor of $2^2$ (4).
(b) $W/1.1^2 = W/1.21 = 0.83W$
**1.10** $64F/9$
mass $\propto$ volume
volume of sphere $\propto$ diameter$^3$
therefore, mass of sphere $\propto$ diameter$^3$.
If the diameter increases by a factor of 2, the mass increases by $2^3$, i.e. 8. Gravitational force is proportional to the product of the mass of the attracting body and the mass of the attracted body, so the force will increase by a factor of $8 \times 8$, i.e. 64. If the distance increases by a factor of 3, the force decreases by $3^2$, i.e. 9. Therefore, the attraction is 64/9 times the force $F$.
**1.11** Estimates: mass of human being 70 kg, distance apart 0.5 m.
Using

$$F = \frac{Gm_1m_2}{d^2}$$

$$F = \frac{6.7 \times 10^{-11} \times 70 \times 70}{(0.5)^2}\ \text{N} \approx 10^{-6}\ \text{N}$$

which is undetectable. It is not worth being precise since human beings are nothing like spheres anyway!

**1.12** (a) $2\Delta$
(b) $\tan 2\theta = \dfrac{\Delta}{D}$

for small angles $2\theta = \dfrac{\Delta}{D}$, $\theta = \dfrac{\Delta}{2D}$

(c) $F = G\dfrac{mM}{d^2}$

(d) $T = \dfrac{GmM}{d^2}\cdot 2b$

**1.19** (a) A positively charged polystyrene sphere would be repelled by a positively charged body and attracted by a negatively charged body.
(b) Test for repulsion.
**1.21** (a) Figure A1 is a graph of $x$ against $1/d^2$.

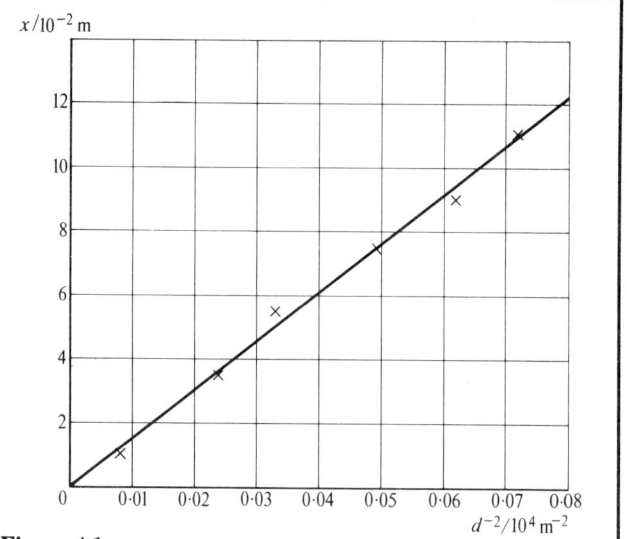


**Figure A1**

(b)
$$F = \frac{kQ_1Q_2}{d^2} \quad \text{and} \quad F = \frac{mgx}{l}$$

Thus $$\frac{mgx}{l} = \frac{kQ_1Q_2}{d^2}$$

rearranging we have

$$x = \frac{kQ_1Q_2l}{mg}\cdot\frac{1}{d^2}$$

Slope of graph $= \dfrac{kQ_1Q_2l}{mg}$

(c) The force constant is of the order $10^{10}$ N $m^2$ $C^{-2}$.

$$k = \frac{\text{slope of graph} \times mg}{Q_1Q_2l}$$

From graph

$$\text{slope} = \frac{12.2 \times 10^{-2}}{0.08 \times 10^4}\ m^3$$

$$k = \frac{12.2 \times 10^{-2}}{0.08 \times 10^4} \times \frac{1.1 \times 10^{-3}}{25 \times 10^{-18} \times 0.5}\ \text{N m}^2\,\text{C}^{-2}$$
$$= 13.42 \times 10^9\ \text{N m}^2\,\text{C}^{-2}$$
$$\approx 10^{10}\ \text{N m}^2\,\text{C}^{-2}$$

**1.22** (a) $2.3 \times 10^{-12}$ N
Using $F = \dfrac{1}{4\pi\varepsilon_0}\cdot\dfrac{Q_1Q_2}{r^2}$ we have

$$F = \frac{9 \times 10^9 \times (1.6 \times 10^{-19})^2}{(10^{-8})^2}\ \text{N}$$

$$F = 2.3 \times 10^{-12}\ \text{N}$$

(b) The force which the electron exerts on the proton is equal and opposite to the force which the proton exerts on the electron.

Acceleration of electron = $2.5 \times 10^{18}\ \text{ms}^{-2}$

Acceleration of proton = $1.4 \times 10^{15}\ \text{ms}^{-2}$

**1.23**

$$\frac{F_E}{F_G} = \frac{1}{4\pi\varepsilon_0} \cdot \frac{Q_1Q_2}{d^2} \div \frac{Gm_1m_2}{d^2}$$

$$= \frac{Q_1Q_2}{G4\pi\varepsilon_0 m_1m_2}$$

$$= \frac{1}{6.7 \times 10^{-11}} \times 9 \times 10^9 \times \frac{1.6 \times 10^{-19} \times 1.6 \times 10^{-19}}{1.7 \times 10^{-27} \times 1.7 \times 10^{-27}}$$

$$\approx 10^{36}$$

The electric force of repulsion between the protons is $10^{36}$ times the gravitational attraction.

## Topic 2

**2.1** (a) Magnetic, gravitational, nuclear.

(b) This will depend on the magnitude of the force and features of the situation. Nuclear forces are very short range within the atomic nucleus. Gravitational forces extend throughout the universe.

(c) Shielding is used to limit magnetic fields.

**2.2** (a) In the opposite direction.

(b) The force would again be in the opposite direction.

**2.3** (a) A charge with small dimensions does not distort the field unduly. A small magnitude of charge avoids induction effects on nearby conductors.

(b) The determining factors are: the magnitudes of the charge on the sphere and the small charge, and their distance apart. (It will also depend on the medium in which they are placed. This is considered in the unit *Electronic systems*.)

**2.5** (a) $4.0 \times 10^5\ \text{N C}^{-1}$

(b) $8.0 \times 10^{-15}$ N

**2.7** (a) Work done is force × distance

$W = Fd$ and $F = EQ$

(b) From the definition of the volt

$W = VQ$

(c) $EQd = VQ$

Thus

$$E = \frac{V}{d}$$

**2.9** Gravitational field strength $g$ is defined as force per unit mass, that is $g = F/m$. If a body is falling freely the force of gravity will cause it to have an acceleration $g$. (Hence from Newton's second law $g = F/m$.)

**2.10** (a) 160 N

(b) 1200 N

**2.12** (a) In repulsion the force is in the positive direction, that is, away from the 'source charge' at the origin. In attraction, the force is in the negative direction.

(b) (i) $50\ \text{N C}^{-1}$ towards B

(ii) $50\ \text{N C}^{-1}$ towards A

(iii) $150\ \text{N C}^{-1}$ towards B

(iv) $50\ \text{N C}^{-1}$ towards A

**2.13** (a) $3.6 \times 10^{10}\ \text{N C}^{-1}$

(b) $5.8 \times 10^{-9}$ N

**2.15** (a) $g_p/4$

(b) $g_p/10^4$

**2.16** (a) The gravitational attraction exerted by the earth on the spaceship will be equal and opposite to the gravitational attraction exerted by the mass on the spaceship.

(b) $$\frac{GM_Em}{r_1^2} = \frac{GM_Mm}{r_2^2}$$

where $m$ is the mass of the spaceship.

Hence $$\frac{r_1^2}{r_2^2} = \frac{M_E}{M_M}$$

(c) $1.6\ \text{N kg}^{-1}$

Consider an object of mass $m$ on the surface of the moon

$$mg_M = \frac{GmM_M}{r^2}$$

where $M_M$ is the mass of the moon, $g_M$ is the moon's gravitational field strength at its surface, and $r$ is the radius of the moon.

But $$M_M = \frac{4\pi}{3} r^3\rho$$

where $\rho$ is the density of the moon's material.

Thus

$$g_M = \frac{4}{3} G\pi\rho r$$

$$g_M = (4\pi/3 \times 1.7 \times 10^6 \times 3.4 \times 10^3 \times 1.7 \times 10^6)\ \text{N kg}^{-1}$$

$$g_M = 1.6\ \text{N kg}^{-1}$$

**2.17** Basic assumptions: (i) A negative mass would attract a negative mass and repel a positive mass. Like masses attract; unlike masses repel (compare with like charges repel, unlike charges attract). (ii) A sphere of negative mass behaves as if all its mass was at its centre. (iii) The gravitational field of a point negative mass varies inversely as the square of the distance from the point.

(a) A positive mass would experience a repulsive force in this gravitational field (compare with the electric field around a positively charged sphere). The lines of force would be directed radially away from the (spherical) object. A spaceship (of ordinary mass) would easily escape from this field! (But could it enter the field in the first place?) (b) It would move in an elliptical orbit. The gravitational force between the two negative masses is attractive.

(c) It would move in a hyperbolic orbit. The planet would experience a repulsive force. This is rather like the path taken by alpha particles when fired at a gold leaf target (the scattering experiment).

**2.19** The work done when the point of application of a force moves is defined as the product of the force and the distance this point moves in the direction of the force.

**2.20** Gravitational force on a mass $m$ is $mg$.

(a) To lift the object through a vertical height of $h$ (without a change of speed) an equal and opposite force must be applied.

Now $W = Fh$

Hence $W = mgh$

(b) Chemical energy to gravitational potential energy.

**2.21** (a) 100 J

(b) 100 J

(c) 100 J

**2.22** $20\ \text{J kg}^{-1}$

**2.23** (a) The work done depends upon (i) the magnitude of the charge, (ii) the electric field strength, (iii) the distance moved.

(b) Electric force on a charge $Q_0$ is $EQ_0$. To move the charge through a distance $x$ an equal and opposite force must be applied.

**2.24** (a) $1.0 \times 10^{-6}$ J
(b) $1.0 \times 10^{-6}$ J
(c) 0.5 J $C^{-1}$
**2.26** (a) The potential at P is 3000 V.
(b) The p.d. between P and Q is 2000 V.
The potential at Q is 1000 V.
(c) The work done is $3.0 \times 10^{-4}$ J.
**2.28** No work is done; all points on an equipotential surface are at the same potential.
**2.29** Spherical
**2.30** The field is represented by field lines, in terms of field strength which enables the force on a charge or mass at a point in the field to be calculated. The field is represented by equipotential lines or surfaces, which enable the potential energy of a charge or mass at a point in the field to be calculated. Field strength is a vector quantity whereas potential is a scalar quantity.

**2.32** $1 \times \text{V m}^{-1} = 1 \times \text{J C}^{-1}\text{ m}^{-1}$
$= 1 \times \text{N C}^{-1}$
$(1 \times \text{N} = 1 \times \text{J m}^{-1})$
**2.33** (a) A smaller angle. The electric field strength between the plates is reduced hence the force exerted on the charged strip is less.
(b) Initially, $5.0 \times 10^4$ V $m^{-1}$ and finally $2.5 \times 10^4$ V $m^{-1}$.
(c) 5000 V
**2.34** (a) See Figure A2.

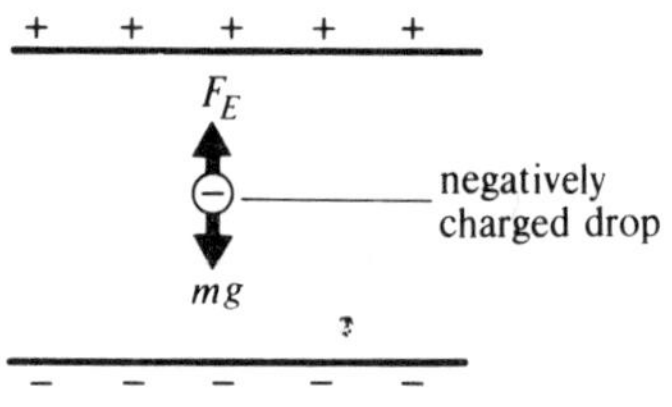


$F_E$ is the force due to the charged plates
$mg$ is the Earth's gravitational pull

**Figure A2**

(b) $2.0 \times 10^{-12}$ C

$$F_E = mg$$

But $$F_E = EQ = \frac{VQ}{d}$$

Hence $$\frac{VQ}{d} = mg$$

or $$Q = \frac{mgd}{V}$$

$$Q = \frac{(10^{-7} \times 10 \times 20 \times 10^{-3})}{10^4}\ \text{C}$$

$$Q = 2.0 \times 10^{-12}\ \text{C}$$

(c) 20 V
The resultant force on the charged drop is $2 \times 10^{-9}$ N upwards (from $F = ma$).
The resultant force is equal to the increase in the electrical force

$$\frac{(V \times 2 \times 10^{-12})}{20 \times 10^{-3}\ \text{m}}\ \text{C}$$

where $V$ is the increase in potential difference.
**2.35** (a) 18 000 eV
(b) $8.0 \times 10^7$ m $s^{-1}$

The work done on the electron by the electric field is equal to the gain in kinetic energy.

Hence $$eV = \frac{1}{2}mv^2$$

or $$v = \sqrt{\left(\frac{2eV}{m}\right)}$$

$$v = \sqrt{(2 \times 1.8 \times 10^{11} \times 18\,000)}\ \text{m s}^{-1}$$

$$v = 8.0 \times 10^7\ \text{m s}^{-1}$$

**2.37** (a) A negative charge would appear on the probe (by induction).
(b) A positive charge would appear on the electroscope.
(c) Positive and negative ions.
(d) The positive ions are attracted to the probe and will neutralise its charge.
**2.38** (a) 3.0 kV
(b) 2.0 kV
(c) 1.5 kV

**2.39** (a) $$F = \frac{1}{4\pi\varepsilon_0}\cdot\frac{QQ_0}{x^2}$$

(c) This area represents the work done by the external agent in moving the test charge through a distance $\Delta x$.
(d) The area between the $x$-axis and the curve between $x = r$ and infinity. It can be shown by integration that the work done $W$ is given by

$$W = \frac{1}{4\pi\varepsilon_0}\cdot\frac{QQ_0}{r}$$

**2.40** (a) $$V = \frac{1}{4\pi\varepsilon_0}\frac{Q}{r}$$

$$\frac{1}{4\pi\varepsilon_0} = \frac{Vr}{Q}$$

Expressing this in form of units we have

$$\frac{1}{4\pi\varepsilon_0} = \frac{Vm}{C}$$

Now $$\text{V} = \frac{\text{J}}{\text{C}} \quad \text{and} \quad \text{J} = \text{N m}$$

Thus $$\frac{1}{4\pi\varepsilon_0} = \frac{\text{N m}}{\text{C}}\cdot\frac{\text{m}}{\text{C}} = \text{N m}^2\ \text{C}^{-2}$$

(b) $$\frac{1}{4\pi\varepsilon_0} = 9 \times 10^9$$

$$\varepsilon_0 = \frac{1 \times 10^{-9}}{36\pi} = 8.9 \times 10^{-9}\ \text{N}^{-1}\ \text{m}^{-2}\ \text{C}^2$$

**2.41** $4.5 \times 10^3$ V
This example illustrates the fact that large potentials arise in the electric field of a comparatively small static charge.
**2.42** (a) 29 V
(b) −29 eV or $-4.6 \times 10^{-18}$ J
(c) $4.6 \times 10^{-18}$ J
(d) $$\frac{mv^2}{r} = \frac{1}{4\pi\varepsilon_0}\cdot\frac{e^2}{r^2}$$
(f) $2.3 \times 10^{-18}$ J
(g) $-2.3 \times 10^{-18}$ J

**2.43** (a) The charge in kinetic energy per kilogram is $-34.16 \times 10^6$ J $kg^{-1}$. The spacecraft is slowing down due to the gravitational pull of the earth. The potential energy of the spacecraft is increasing.
(b) The change in gravitational potential energy per kilogram is $+34.16 \times 10^6$ J $kg^{-1}$.
The spacecraft is now approaching the earth and is being accelerated by the gravitational pull of the earth. The potential energy of the spacecraft is decreasing. The change in gravitational potential energy is $-37.79 \times 10^6$ J $kg^{-1}$.

(c) The values are almost the same. If the distances had been equal it seems likely that the changes in the energy would have been the same.
(d) The gravitational potential difference.
(e) $+20.89 \times 10^6$ kg$^{-1}$.

**2.44** (a) $F = \frac{GMm}{r^2}$

where $G$ is the gravitational constant, $M$ is the mass of the earth and $m$ is the mass of the spacecraft.
(b) The gain in gravitational potential energy per kilogram is equal to the work done per kilogram against the gravitational force and is given by

$$\Delta W = -\frac{gM}{r^2}\cdot\Delta r$$

(c) The gain in potential energy per kilogram is equal to the area between the $x$-axis and the curve, between the values of $r$ equal to $r_1$ and $r_2$. This area can be found by integration.
**2.45** $+26.1 \times 10^6$ J kg$^{-1}$. This is in good agreement with the value obtained from the motion of the spacecraft.
**2.46** (a) $+34.41 \times 10^6$ J kg$^{-1}$
(b) Yes. At the distance $r_0$ all the kinetic energy has been transformed into gravitational potential energy. This is the farthest distance which could be reached without extra energy.
(c) $+35.33 \times 10^6$ J kg$^{-1}$
(d) $+14.44 \times 10^6$ J kg$^{-1}$
(e)

$$-\frac{GM}{r_0} - \left(-\frac{GM}{r}\right)$$

or

$$\frac{GM}{r} - \frac{GM}{r_0}$$

(f) See figure A3.

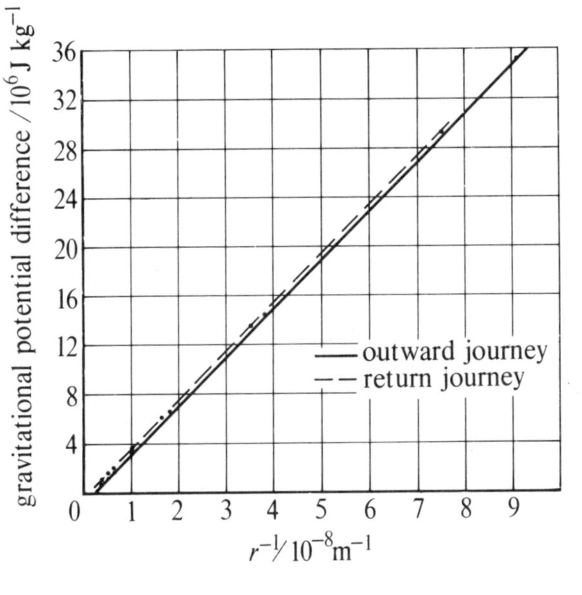


**Figure A3**

(g) Both the graphs should be straight lines. This shows that the gravitational potential difference is proportional to $1/r$.

(h) The slope of each graph is $GM$.
(i) The intercept is $1/r_0$. It is not exactly the same.
**2.47** (b) The negative sign tells us that all points in a field where the forces are attractive have a negative potential. (Forces of repulsion give rise to positive potentials.) This arises because we decided that gravitational potential is zero at infinity and when we go from infinity to a point in the field there is a decrease in potential. (Alternatively, as the distance from the mass increases the potential increases, that is, it gets less negative.)
**2.48** (a) $-62.5 \times 10^6$ J kg$^{-1}$
(b) A rocket must have $62.5 \times 10^6$ J kg$^{-1}$ if it is to escape from the earth's gravitational field.

**2.49** (a) $V = \frac{1}{4\pi\varepsilon_0}\cdot\frac{Q}{r}$

(b) The potential is negative.
**2.53** 30 V

## Topic 3

**3.2** Forces due to the vertical components of the field will produce a torque in a vertical plane, but the compass cannot rotate in a vertical plane.
**3.3** (a) (i) The direction of the line, or tangent to it at any point, is the direction of the force which would act on a north seeking pole at that point, i.e. it is the direction in which a small compass sets when freely suspended through its centre of gravity. (ii) A position in a magnetic field where the resultant field is zero.
(b) (i) For a bar magnet a field line is shown starting at one surface and ending at another. But that is because we have not shown the field inside the magnet. For a solenoid the magnetic field is shown as a closed loop—does that have a beginning or an end?
(ii) No, they do not cross—they do sometimes seem incomplete because they run off the paper, however, it is reasonable to suppose that however far they go into space and however weak they become they are continuous from pole to pole.
(iii) By parallel lines.
(iv) Yes the plotting compass could be used to investigate the direction of the magnetic field of the small bar magnet. The field of the compass will be symmetrical and hence the direction of field is unaffected.
**3.4** Maxwell' screw rule. If a *single handed* screw is turned so

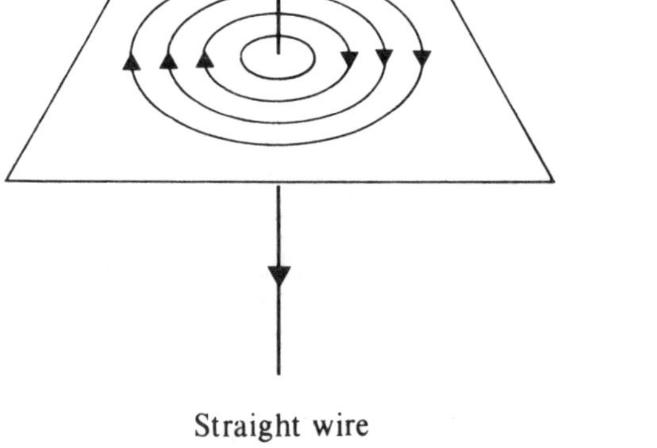


**Figure A4a**

that it moves *forward* in the *direction* of the *current*, then the *direction* of *rotation* gives the *direction* of the *magnetic field*. Right-hand grip rule. If a conductor carrying a current is gripped with the *right hand* with the *thumb* pointing in the *direction* of the *current*, the *fingers* will point in the *direction* of the *magnetic field*.

The magnetic field pattern due to a current in a straight conductor is given in figure A4.

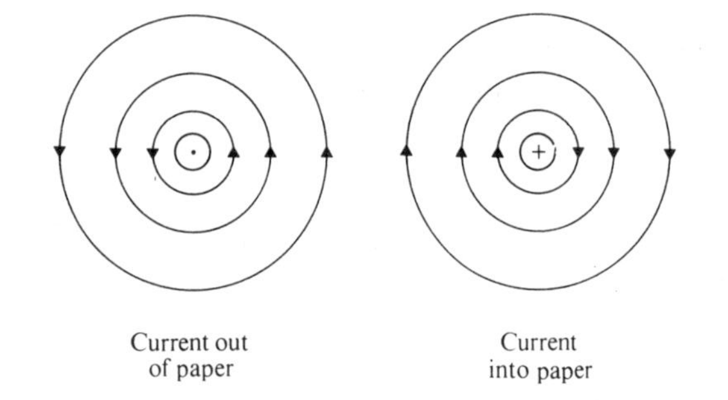


**Figure A4b**

**3.5** (a) See figure A5.

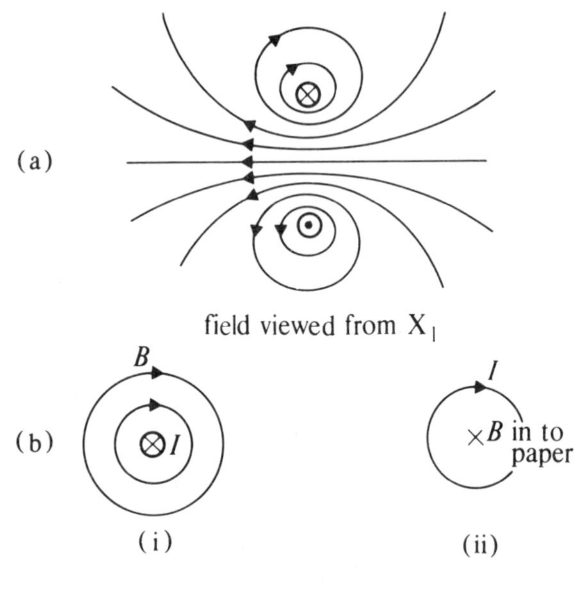


**Figure A5**

(b) Yes. The figures represent: (i) the field around the right-hand side of the loop; (ii) the field along the axis of the loop viewed from $Y_2$.

**3.7** See figure A6.

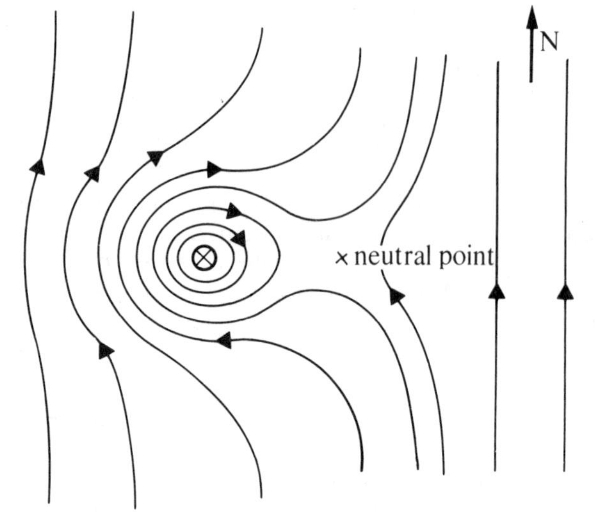


**Figure A6**

**3.9** (a) and (d)

**3.10** See figure A7.

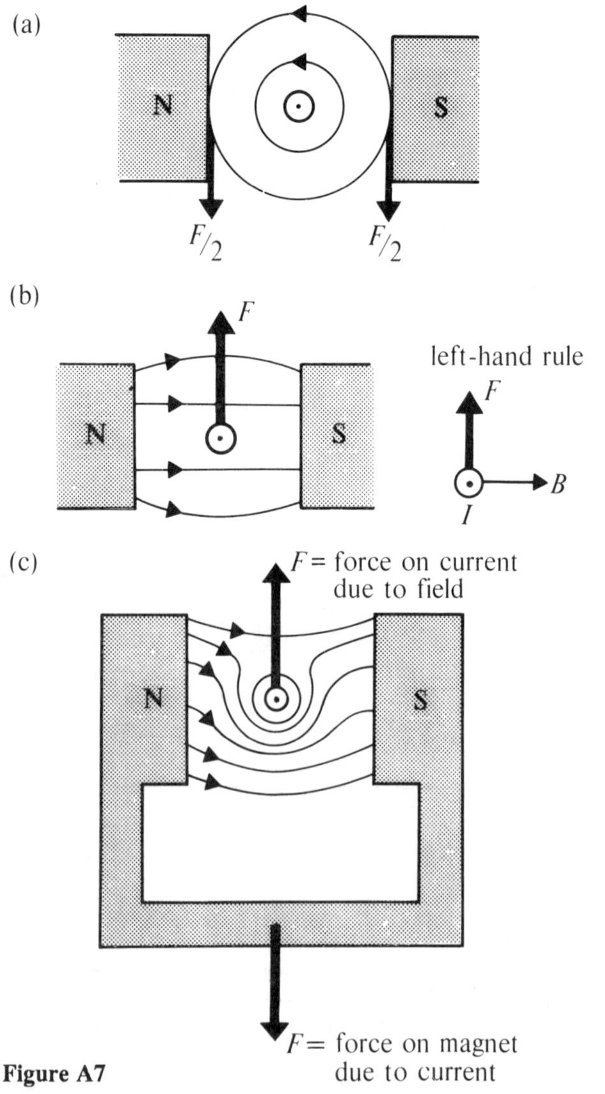


**Figure A7**

(c) Towards a weaker field.

**3.11** Both segments A and B will be acted on by a magnetic force away from the axis of the solenoid (along a radius). A sudden high current would tend to make the coils of the solenoid expand away from the axis and possibly 'explode'.

**3.12** $1\text{T} = 1\text{ N A}^{-1}\text{ m}^{-1}$

$$[F] = [\text{M L T}^{-2}]$$
$$[I] = [\text{A}]$$
$$[l] = [\text{L}]$$
$$[T] = \frac{[\text{M L T}^{-2}]}{[\text{A}][\text{L}]} = [\text{M A}^{-1}\text{ T}^{-2}]$$

1 tesla is the magnetic flux density which produces a force of 1 newton on a conductor of length 1 metre, carrying a current of 1 ampere, placed at right angles to the magnetic field.

**3.13** (a) (i) $B = \dfrac{1.2 \times 9.8 \times 10^{-4}\text{ N}}{1.4\text{ A} \times 0.25\text{ m}}$

$= 3.4 \times 10^{-3}$ T for current 1.4 A

(ii) $2.4 \times 10^{-3}$ T for 1 A

(b) $I = 7.5$ mA $\quad I = \dfrac{1.8 \times 10^{-5}}{2.4 \times 10^{-3}}$

**3.14** (a) $l = vt$

(b) $NQ$

(c) $NQ/t$

(d) $F = \dfrac{BNQl}{t} = \dfrac{BNQvt}{t} = BNQv$

(e) $BQv$

(f) R → L

(g) Out of the paper.

**3.15** (a) Towards the front edge of the slab.

(b) $EQ = BQv, \quad E = Bv$

(c) $E = V_H/d$

(d) $V_H = Bvd$

(e) $V_H = I/nQA = I/nQtd$

$$V_H = \frac{BId}{nQtd} = \frac{BI}{nQt}$$

**3.17** (a) $n$ = free electrons per unit volume $= 1.1 \times 10^{29}\text{ m}^{-1}$

$$n = \frac{BI}{V_H Qt}$$
$$= \frac{0.75\text{T}}{3.8 \times 10^{-6}\text{ V} \times 1.6 \times 10^{-19}\text{ C} \times 10^{-4}\text{ m}}$$

(b) Volume of slice $= 10^{-7}\text{ m}^3$

Volume of 1 mole $= \dfrac{64}{0.89} \times 10^{-7}\text{ m}^3$

$7.2 \times 10^{-6}\text{ m}^3$ contains $6 \times 10^{23}$ atoms

$1\text{ m}^3$ contains $8.3 \times 10^{28}$ atoms.

(c) 1.3 free electrons per atom

(d) $v = \dfrac{1}{nQA}$

$$= \frac{9.0\text{A}}{1.1 \times 10^{29}\text{ m}^{-3}} \times \frac{1}{1.6 \times 10^{-19}\text{ C} \times 2 \times 10^{-16}\text{ m}^2}$$
$$= 2.6 \times 10^{-4}\text{ m s}^{-1}$$

**3.18** $$B = \mu_0 \frac{NI}{l}$$

Rearranging we have

$$\mu_0 = \frac{Bl}{NI}$$

unit for $\mu_0$ is therefore $\dfrac{\text{T m}}{\text{A}}$ or T m A$^{-1}$

Now $\quad 1\text{ T} = 1\text{ N A}^{-1}\text{ m}^{-1}$

Substituting for T we have

$$\text{N A}^{-1}\text{ m}^{-1}\text{ m A}^{-1} = \text{N A}^{-2}$$

**3.19** (a) $1.0 \times 10^{-2}$ T

(b) $5.0 \times 10^{-3}$ T

(c) $1.0 \times 10^{-4}$ T

(d) $6.3 \times 10^{-3}$ T

**3.20** (a) None

(b) $15.7 \times 10^{-6}$ T

**3.21** (a) Axis of coil in magnetic meridian. Plane of coil is 70° to vertical (see figure A8).

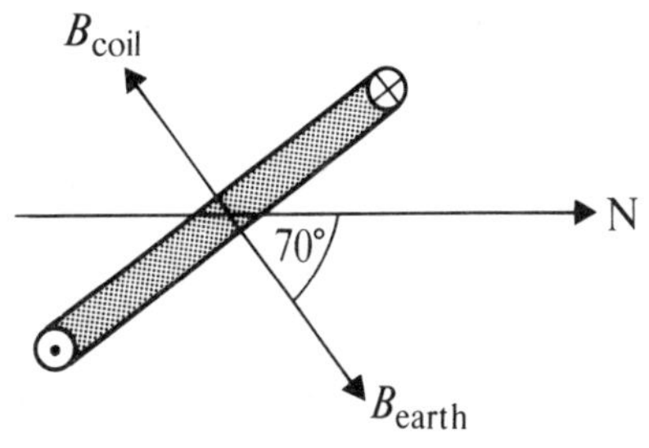


**Figure A8**

(b) Using $\quad B = \dfrac{\mu_0 NI}{2a}$

$$5.2 \times 10^{-5}\text{ T} = \frac{4\pi \times 10^{-7}\text{ N A}^{-2} \times 160 \times I}{2 \times 0.60\text{ m}}$$
$$I = 0.31\text{ A}$$

**3.22** (a) $B = \dfrac{\mu_0 I_1}{2\pi a}$

(b) $F = \dfrac{\mu_0 I_1 I_2 l}{2\pi a}$

(c) (i) The current-carrying conductors are interacting so force $F$ is the reaction force on $I_1$ due to force $F$ acting on $I_2$ (a 'third-law pair').

(ii) $-F$ is also the force on $I_1$ due to the magnetic field created by $I_2$ along the path of $I_1$ and its direction is given by Fleming's left-hand rule.

(d) $F = \dfrac{4\pi \times 10^{-7}}{2\pi}\text{ N} = 2\pi \times 10^{-7}\text{ N}$

**3.25** $\quad [c] = [\text{L T}^{-1}]$

Unit of $\mu_0$ is N A$^{-2}$

$$[\mu_0] = [\text{M L T}^{-2}][\text{A}^{-2}] = [\text{M L T}^{-2}\text{ A}^{-2}]$$

Unit of $\varepsilon_0$ is C$^2$ N$^{-1}$ m$^{-2}$

$$[\varepsilon_0] = [\text{A}^2\text{ T}^2][\text{M}^{-1}\text{ L}^{-1}\text{ T}^2][\text{L}^{-2}] = [\text{M}^{-1}\text{ L}^{-3}\text{ T}^4\text{ A}^2]$$

Thus $\quad [\mu_0\varepsilon_0] = [\text{M L T}^{-2}\text{ A}^{-2}][\text{M}^{-1}\text{ L}^{-3}\text{ T}^4\text{ A}^2] = [\text{L}^{-2}\text{ T}^2]$

Therefore

$$\frac{1}{[\mu_0\varepsilon_0]^{1/2}} = [\text{L T}^{-1}]$$

**3.26** (a) See figure A9.

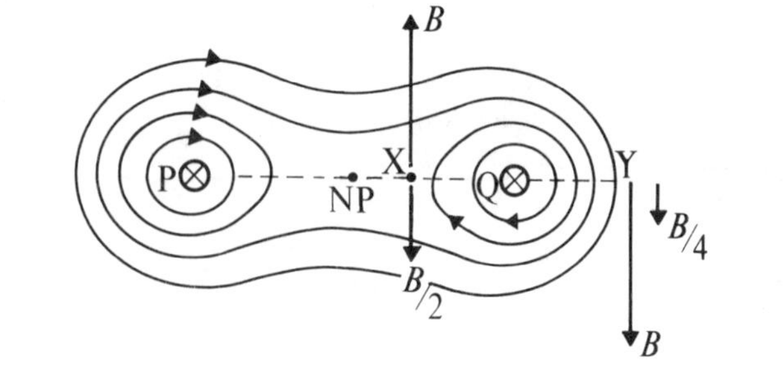


**Figure A9**

(b) Field at X had flux density $B/2$, at Y $5B/4$. Forces acting between currents in the same direction are attractive and tend to move the wires towards the weaker field between the wires.

(c) See figure A10. The $B$-field at Q due to current in P produces a force $F$ on Q away from P and a force $-F$ acts on P away from Q.

The force tends to move the wires away from the region of strong magnetic field between the wires.

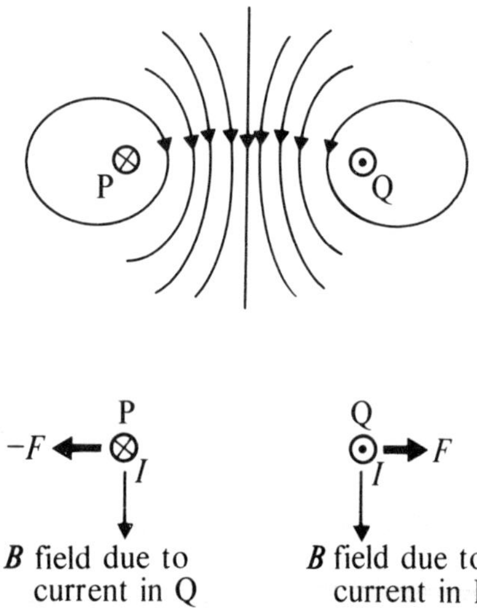


**Figure A10**

**3.27** (a) See figure A11.

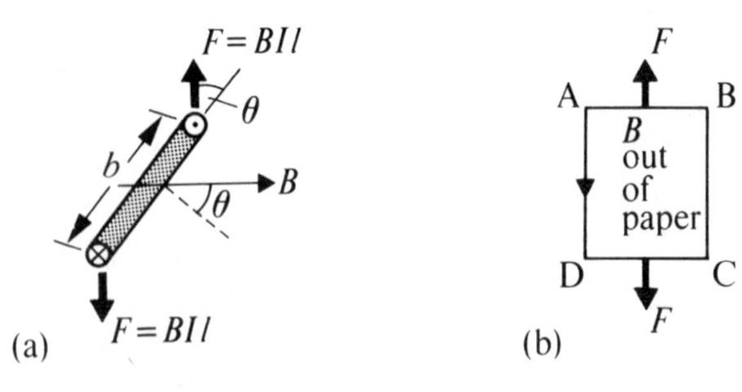


**Figure A11**

(b) Yes; a force acts on AB out of the plane of the paper. An equal and opposite force on CD along the same line. There is no resultant effect—no resultant force, no resultant moment.

(c) Perpendicular distance between forces on AD and BC is $b \sin\theta$.

Torque $= BIlb \sin\theta = BIA \sin\theta$ for 1 turn.

Torque for $N$ turns $= BIAN \sin\theta$.

Torque is maximum when $\theta = 90°$ i.e. plane of coil parallel to B.

**3.30** (a) $\left(3.6 \times 10^{-4} \dfrac{\text{N m}}{\text{A}}\right) I$;

$T = NAIB$

$= 100 \times 2.4 \times 10^{-4}\ \text{m}^2 \times 1.5 \times 10^2\ \text{T} \times I$

(b) $9 \times 10^{-7}$ N m

(c) 2.5 mA

(d) 0.2 rad per mA

(e) 8 mm per mA

## Topic 4

**4.1** (a) $GM\left(\dfrac{1}{R} - \dfrac{1}{r}\right)$; $\quad -\dfrac{GM}{r} - \left(-\dfrac{GM}{R}\right)$

(b) (i) $GMm\left(\dfrac{1}{R} - \dfrac{1}{r}\right)$; (ii) $\dfrac{GMm}{R}$

(c) $$\frac{1}{2} m v_e^2 = \frac{GMm}{R}$$

therefore $$v_e^2 = \frac{2GM}{R}$$

and $$v_e = \sqrt{\left(\frac{2GM}{R}\right)}$$

(d) $v_e$ for earth $= 1.1 \times 10^4\ \text{m s}^{-1}$

$v_e$ for moon $= 2.3 \times 10^3\ \text{m s}^{-1}$

**4.2** (a) The r.m.s. speed of hydrogen molecules is about 1/5 of the escape velocity from the earth at 300 K. Even at this temperature, a significant proportion of $H_2$ molecules would be moving fast enough to escape and the earth could not retain much hydrogen. The temperature in the upper atmosphere is much greater than 300 K, hence it is impossible for the earth to keep hydrogen in the atmosphere.

(b) The r.m.s. speed of oxygen molecules is 1/30 the escape speed from the earth. This enables the earth to retain oxygen molecules in the atmosphere. The r.m.s. speeds of oxygen and hydrogen molecules at 300 K is of the same order as the escape speed from the moon—hence no atmosphere is retained.

**4.3** (a) Because the force acting is always perpendicular to the velocity of the body and hence produces no change in speed.

(b) Gravitational force.

(c) $$\frac{GMm}{r^2} = \frac{mv_0^2}{r}$$

then $$v_0^2 = \frac{GM}{r} \quad \text{and} \quad v_0 = \sqrt{\left(\frac{GM}{r}\right)}$$

(d) $$v_0 = \frac{2\pi r}{T}, \quad T = \frac{2\pi r}{v_0}$$

$$T^2 = \frac{4\pi^2 r^2}{v_0^2}, \quad = 4\pi^2 r^2 \cdot \frac{r}{GM}$$

**4.4** (a) $r = 6.4 \times 10^6\ \text{m} + 2 \times 10^4\ \text{m}$

$= 6.6 \times 10^6\ \text{m}$

$$v_0 = \frac{GM}{r} = 7.8 \times 10^3\ \text{m s}^{-1}$$

(b) $$T = \frac{2\pi r}{v_0} = \frac{2\pi \times 6.6 \times 10^6}{7.8 \times 10^3}\ \text{s}$$

$= 5.3 \times 10^3$ s

$= 88$ mins

(c) The first three satellites listed all have heights about 200 km and orbit times of the order of 90 mins.

**4.5** (a) $\dfrac{10}{60 \times 60}$ N kg$^{-1}$ since $F \propto \dfrac{1}{r^2}$

(b)
$$\frac{v^2}{60R} = \frac{10}{60 \times 60}$$
$$v = \sqrt{\left(\frac{6.4 \times 10^6}{6}\right)}\ \text{m s}^{-1}$$
$$= 1.0 \times 10^3\ \text{m s}^{-1}$$
$$\text{Period} = \frac{2\pi \times 60 \times 6.4 \times 10^6}{10^3}$$
$$= 2.4 \times 10^6\ \text{s} = 28\ \text{days}$$

**4.6** For any satellite of the earth
$$T^2 = \frac{4\pi^2 r^3}{GM}$$
where $M$ is the mass of the earth

Using moon's orbit data
$$M = \frac{4\pi^2 \times (3.8 \times 10^8)^3\ \text{m}^3}{6.7 \times 10^{-11}\ \text{N m}^2\ \text{kg}^{-2} \times (2.4 \times 10^6)^2\ \text{s}^2}$$
$$= 6.8 \times 10^{24}\ \text{kg}$$

**4.7** (a) The communication satellite has an orbit period of 1 day, therefore
$$\frac{1}{(27.7)^2} = \frac{x^3R^3}{60^3R^3}$$
$$x^3 = \frac{60^3}{(27.7)^2} = 280, \quad x = 6.5$$

(b) The satellite is 5.5 $R$ above the earth's surface.

Height = 35 000 km

Yes, the Early Bird satellite is a communication satellite with an orbit height of 35 000 km.

**4.8** (a) If the neutral point is distance $r_1$ from the earth's centre and $r_2$ from the moon's centre
$$\frac{GM}{r_1^2} = \frac{GM_m}{r_2^2}$$
$$\frac{r_1}{r_2} = \sqrt{\left(\frac{M}{M_m}\right)} = \sqrt{\left(\frac{6.0 \times 10^{24}}{7 \times 10^{22}}\right)} = \sqrt{\left(\frac{600}{7}\right)} = \frac{9.2}{1}$$
$$\frac{r_2}{r_1 + r_2} = \frac{1}{10.2}$$
$r_2$ is 9.8% of total distance between centres.

(b) Near A the craft is losing speed (has negative acceleration) in the moon's gravitational field.

At B the speed is constant because the craft is passing through a region where the resultant gravitational field is zero.

At C the craft is gaining speed or accelerating in a region where the gravitational field is towards the earth.

Yes; the answer to (a) predicts a neutral point, 90.2% of 3.8 $\times 10^8$ m from the earth, i.e. at $3.4 \times 10^8$ m from the earth.

**4.9** Work done on the electron by the electric field is equal to the increase in kinetic energy.
$$eV = \frac{1}{2} m_e V^2$$
Thus
$$v = \sqrt{\left(\frac{2eV}{m_e}\right)}$$

**4.10** (a) $10^4$ V m$^{-1}$, from A to C.

(b) $1.6 \times 10^{-15}$ N, from C to A.

(c) $8.0 \times 10^{-17}$ J

(d) $1.3 \times 10^7$ m s$^{-1}$

(e) The same.

**4.11** (a) P is negative with respect to Q.

(b) Horizontal velocity constant.

(c) There is a constant vertical force and uniform vertical acceleration.

(d) A straight line, but see answer to (e).

(e) Since the electron has mass it will move in a parabola in the earth's gravitational field, but if the electron's horizontal speed is very large it will appear to travel in a straight line.

(f) In a field of $10^4$ V m$^{-1}$ the force on an electron is $1.6 \times 10^{-15}$ N. By contrast the gravitational force is $mg$ which is only $9 \times 10^{-30}$ N, a difference in order of magnitude of more than $10^{14}$. Hence, gravitational effects can be ignored.

**4.12** (a) $x = vt$

(b) Acceleration $a = \dfrac{Ee}{m_e}$
$$y = \frac{1}{2}at^2 = \frac{1}{2}\frac{Ee}{m_e}\cdot t^2$$

(c) $y = \dfrac{1}{2}\dfrac{Ee}{m_e}\cdot\dfrac{x^2}{v^2}$

(d) (i) Uniform electric field. (ii) No other forces acting on the electron.

**4.13** (a) The force is always perpendicular to the direction of its instantaneous velocity.

(b) Magnetic force = $Bev$, and this is constant in size if speed $v$ and flux density $B$ are constant.

(c)
$$\frac{m_e v^2}{r} = \text{Bev}, \quad r = \frac{m_e v}{Be}$$

(d)
$$v = \sqrt{\left(\frac{2eV}{m_e}\right)}$$
therefore, substituting
$$r = \left[\frac{m_e\sqrt{\left(\frac{2eV}{m_e}\right)}}{Be}\right]$$
which gives
$$\frac{e}{m_e} = \frac{2V}{B^2r^2}$$

**4.14** $B = 9.1 \times 10^{-4}$ T

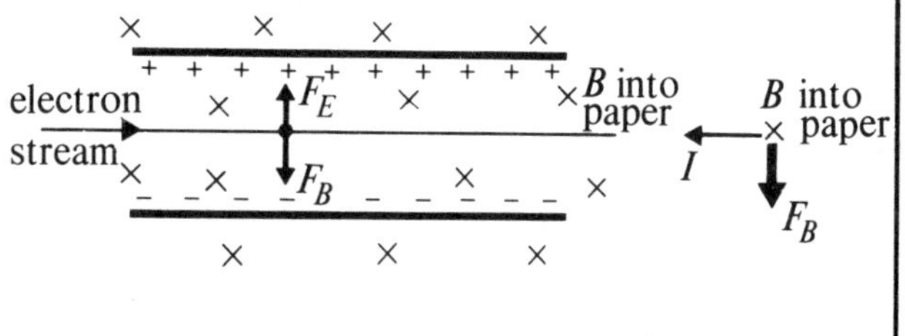


**Figure A12**

**4.15** (a) See figure A12.

(b) $Ee = Bev$

(c) $10^4$ m s$^{-1}$

**4.16** (a) The orbit radius increases.

(b) $T = \dfrac{2\pi r}{v} = \dfrac{2\pi mv}{Bev} = \dfrac{2\pi m}{Be}$

(c) $T$ is independent of $v$ and so remains constant for successive orbits. (This assumes that $m$ is constant and this is not the case for high energy electrons travelling at speeds approaching $c$.)

(d) Frequency $= \dfrac{Be}{2\pi m} = 1.2 \times 10^7$ s$^{-1}$

**4.17** (a) An electron at A experiences a force perpendicular to the equipotential which gives it an acceleration from left to right and also an acceleration towards the axis.

(b) (i) It will tend to converge, (ii) it will converge less or diverge because there is an acceleration component away from the axis.
(c) High speed electrons will experience less sideways deflection than slower speed electrons.

**4.18** (a)

$$\frac{1}{2}mv^2 = eV, \quad v = \sqrt{\left(\frac{2eV}{m}\right)}$$

$$\approx 5 \times 10^7 \text{ m s}^{-1}$$

$$\lambda = \frac{6.63 \times 10^{-34}}{9.1 \times 10^{-31} \times 5 \times 10^7}$$

$$\approx 10^{-2} \text{ nm}$$

(b)

$$\text{Ratio } \frac{\lambda_{\text{electrons}}}{\lambda_{\text{light}}} \approx \frac{10^{-2}}{500} \approx 2 \times 10^{-5}$$

(c)

$$\lambda_{\text{He}} \approx 10^{-4} \text{ nm}$$

$$\text{Ratio } \frac{\lambda_{\text{He}}}{r_{\text{Cu}}} \approx \frac{10^{-4}}{3 \times 10^{-9}} \approx 3 \times 10^4$$

# Index

## References

These textbooks may be reffered to at the start of each topic. Full details are to be found in the *Student's Resource book*


| | |
|---|---|
| Akrill | Akrill,T.B. Bennet, G.A.G. and Miller, C.J. *Physics.*<br>Edward Arnold |
| Bolton | Bolton, T. *Patterns in Physics.*<br>McGraw-Hill |
| Duncan | Duncan,T.*Physics: A text book for advanced level students.*<br>John Murray |
| Muncaster | Muncaster,R. *A-level Physics.*<br>Stanley Thornes |
| Nelkon | Nelkon,M. and Parker, P. *Advanced Level Physics.*<br>Heinemann |
| Wenham | Wenham, F.J. and others. *Physics; Concepts and Models.*<br>Longman |
| Whelan | Whelan, P.M. and Hodgson, M.J. *Essential Principles of Physics.*<br>John Murray |

## Acknowledgements

Thanks are due to the following, who have kindly permitted the reproduction of copyright photographs:

Fig 1.1, NHPA; Fig 1.8, Mary Evans Picture Library; Fig 2.16, Allan Cash; p.51, Science Photo Library; Fig 4.12, Jeal UK Ltd; Fig 3.2a, Fig 3.3, EDC.

The following photographs are by Martin Sookias and Martin Thornton: Fig 3.21, Fig 3.23, Fig 3.23, Fig 3.24.

The following photographs are by Tony Langham: Cover, Fig 2.20, Fig 2.22, Fig 2.23.

Every effort has been made to contact the copyright holders of the following photographs: p. 2, p.13, Fig 2.2, p. 33, Fig 4.8 b. The authors and publishers regret these omissions and will be please to rectify them in future printings.

Examination questions are reproduced by kind permission of the Joint Matriculation Board.

# The Little Warrior's Guide To Cerebral Palsy

VANESSA A. RAY

**ISBN: 9798850230272**